# A Practical Guide to Television Sound Engineering

# A Practical Guide to Television Sound Engineering

Author
Dennis Baxter

Focal Press
Taylor & Francis Group

NEW YORK AND LONDON

First published 2007
This edition published 2013 by Focal Press
70 Blanchard Road, Suite 402, Burlington, MA 01803

Simultaneously published in the UK
by Focal Press
2 Park Square, Milton Park, Abingdon, Oxon OX14 4RN

*Focal Press is an imprint of the Taylor & Francis Group, an informa business*

**Library of Congress Cataloging-in-Publication Data**
Application submitted

ISBN-13: 978-0-240-80723-2 (pbk)
ISBN-13: 978-0-080-92726-8 (ebk)

## Dedication

This book is dedicated to my wife and son, who were patient while I learned and practiced my art. I am grateful they were! Tremendous respect goes to the many talented sound engineers, technicians and designers who every day get the show on the air. Sometimes against all odds!

# Table of Contents

Introduction     xi

**1**    Introduction to TV Sound Engineering     1

**2**    The Senior Audio in Charge     19

**3**    The Mixing Desk     49

**4**    The Audio Assistant     99

**5**    Communication Systems     119

**6**    Cable—Hooking It All Up     141

**7**    Microphones     163

**8**    Wireless Systems     211

Index     237

# Acknowledgments

Thanks to some of the most talented sound designers and audio experts who spent time sharing their thoughts, expertise and vast amounts of knowledge for this book:

|  | Fred Aldous | Peter Adams |
|  | Denis Ryan | Ron Scalise |
|  | Bob Dixon | Klaus Landsberg |

*Editor:* Dr. Jim Owens

*Reviewers:* Ken Reichel — Larry Estrine
Peter Adams — Andrea Borgato
Alexander Lepges — Stamatia Staikoudi
Ralph Strader

*References:*
www.audio-technica.com
www.calrec.com
www.dolby.com
www.dpamicrophones.com
www.euphonix.com
www.lawo.de
www.sennheiser.com
www.solidstatelogic.com
www.stagetec.com
www.telex.com
Yamaha, *Sound Reinforcement Handbook*
Streicher, Ron and Dooley, Wes "Technique for Stereophonic TV and Movie Sound," *AES*, 1988

*Photo Credits*

Cover Photo: Greg Briggs

Richard Garido
Greg Briggs
Matt Benedict
Fred Aldous
Hiroyuki Kobayashi
Chad Roberts
Jim Owens

Manolo Romero – Olympic Broadcast Services
Audio Technica

# Introduction

Of all the jobs in television, audio is one of the least understood and sometimes one of the least appreciated! Television began as a visual medium and engineers were happy just to have a black-and-white image with crackly, low-fidelity voice. For decades television audio remained fairly primitive, using a mixing console the size of a couch to control just a few microphones. The change began with the rapid growth of the electronics industry after WWII. This not only provided equipment and technology to an emerging music and recording industry, but also high-quality electronic components and speakers were available to the home user.

The recording studios required large mixing consoles with quality electronics and a wide range of processing gear and microphones. When the size, weight and reliability of the electronics became manageable, recording equipment was adopted by television audio engineers and integrated into the studios and outside broadcast facilities. Finally the broadcasters were capable of capturing and delivering a high-fidelity sound mix and new standards for quality and creativity were set by the major networks on special events like the Superbowl, Grammy Awards and Daytona 500.

The television screen has grown to a wide 16 x 9 format and television producers have filled the screens with more visual information than the mind can digest. High-quality speakers and electronic components are readily available and consumers are now spoiled by movies and games that deliver good-quality multidimensional sound. Certainly computers and games have changed the way people experience television and sound.

It has taken years to shake the perception that audio follows video, but this has changed as the responsibilities and contributions of audio have enlarged. While it is the job of the audio engineer to mix a good-sounding program and hopefully enhance the viewing experience of the audience, often the challenge of getting to air is demanding.

Many attempts have been made to give television depth and dimension, from 3D glasses to bigger and bigger screen sizes. But nothing enhances the illusion like good sound. The film boys figured out years ago the significance of good sound to a visual presentation. The age of sound is finally here!

For years television sound was viewed as an engineering function in which the audio engineers were responsible for wiring up the audio flow of the television truck as well as communications and anything else that had a speaker and a microphone. The senior audio engineer usually was the team leader and was responsible for all the engineering set-up as well as ultimately mixing the television broadcast. The engineering set-up can be so consuming that the audio mixer is still preparing the television truck as the associate director is counting down to air. A stressful set-up can often lead to a tense show with unintentional errors.

Audio requires a hefty understanding of sound, electronics and systems engineering. But mixing television sound requires a creative construction of aural architecture plus a little artistic voodoo! If you have two different sound engineers mixing identical content, it will sound different because of their subjective interpretations of sound. It should sound different, but it had better sound good! It takes a special creative talent to construct a good-sounding television show under difficult and subjective conditions. The job responsibilities are varied and demanding, but most importantly the sound mixer has to deliver and promote good sound.

The purpose of this book is to provide the reader with a working knowledge of television sound. These pages contain technical overviews of microphones, mixing consoles, communication systems and connectivity and should be a basis for a continuous educational process. Computerization and digital equipment have given us new technology and tools to manage a massive amount of sound, but you still have to listen. You have to develop "good ears" and a strong focus in order to mix a good show.

I hope this book will encourage and stimulate creativity. Mixing is the creative and the subjective interpretation of an orchestra of sounds—some might say a "wall of sound" that has to be tamed!

If you wish to contact the author, please email dbaxter@charbaxter.com.

# 1 Introduction to TV Sound Engineering

The significance of sound to the visual medium cannot be overestimated. Film producers learned many years ago the impact that sound and music had in the success of a screen production. No one will ever forget *Jaws* and how a simple two-note audio cue made everyone's hair stand up.

As television equipment and television remote production trucks, also known as *outside broadcast* (OB) vans, got bigger and more sophisticated, television producers and directors began to pay more attention to sound. A motivated group of audio technicians rose to the occasion with clever ideas and great-sounding results. Recently, a *USA Today* columnist bestowed audio accolades on some and blistered others about the sound of the weekend's television coverage. Finally, people are paying attention to the sound!

The importance and complexity of audio production in today's broacasting environment is particularly evident with the move from analog to digital. It has taken years to settle on standards and the implementation has been inconsistent, but high-definition television with surround sound is here to stay. (See Figure 1.1.)

Producers like Don Ohlmeyer and director Mike Wells pushed the expectations of audio and have inspired superior productions. FOX Network Director of Sports, David Hill has used sound to differentiate FOX Sports from other sports telecasts. It could be argued that FOX has been the most progressive of all the American Broadcasters in implementing new technology and production techniques.

Television has a legacy of being a live medium. Game and quiz shows and variety programs such as *The Ed Sullivan Show* were a main source of programming during the 1950s and 1960s. They were live and by today's standards fairly easy to produce. (Do you think the Beatles got a sound check?) Ed did not use a lavaliere microphone and there were no wireless microphones, which made the job of the "boom operator" the key sound position.

**Figure 1.1** ESPN sideline position with three announcer headsets. This audio position requires three announcer headsets, a camera operator on the directors communication channel plus stage manager and audio communications gear.

It did not take long for some television producer to figure out that live events such as news, sports and entertainment can fill a lot of air time! The remote broadcast was born, even though the equipment was large and sometimes difficult to keep operating. The sound engineer wore a tie and a pocket protector and operated a mixing desk with eight rotary faders and some relays and vacuum tubes. The networks had large engineering departments. The BBC even had a complete R&D department, which can be credited with work on the first digital-mixing consoles.

Technology and television have a symbiotic relationship that thrives on new ideas. Sports and location production drove the development of superior and smaller cameras with better resolution. Recording and playback devices now have multichannel playback and record, and some can do both at the same time. Television sound has evolved from viewers being happy just to see a picture, to the sound being the most technically and aesthetically differentiating factor of a program.

Certainly the role of audio has evolved as well. Television audio is an art that is reliant on technology to combine and manipulate sound to generate a soundtrack for a picture. New technology such as surround sound has opened many creative possibilities for live sound mixers that were formerly only available to post-production film sound. As television sound equipment grows more complex, those in the field must continuously learn in order to survive. If you can master the technology, you can master the art of sound. Dynamic dimensional soundscapes are possible for the creative sound artist.

The senior audio role is to manage the sound responsibilities of the entire production. This includes show sound in all audio formats, managing the audio team, organizing and programming the communications, routing and distributing the audio elements and transmitting and sometimes troubleshooting a problem while on-the-air. The sound mixer/designer role is to create a full rich soundscape in real time. Live! Most multicamera shoots, entertainment, sports or informational programs are live to a viewing audience or live to tape.

There is a tremendous amount of accountability beyond "mixing a good show." A growing trend is toward having an audio technician travel with the television truck because of the variety of digital consoles, routers and programming involved in getting to air. For complex productions, the set-up has grown tremendously, in addition to all the various mixes, outputs and formats that the audio mixer has to generate, plus mix a good-sounding show. If a piece of equipment has a microphone, speaker or audio circuit, it is the responsibility of the audio department—and it's a lot of work! If you ask most people in the field why they love it, a standard and probably honest answer is "the challenge of doing it live"! No second take, and no taking it off the airwaves! In television, you are only as good as your last show, and fortunes and misfortune turn fast and sometimes unfairly.

**Figure 1.2** The mixing console is the control center for all audio signals.

Televised sports exploded during the 1980s. Rights fees were cheap and sports were the mainstay of cable programming. This led to rapid growth at ESPN, which prompted the rise of cable TV and the concept of a nationally distributed cable station like the Superstation Channel 17 and WGN Chicago. The Atlanta Braves baseball franchise provided a lot of hours of airtime when Channel 17 and Ted Turner were pioneering cable television programming.

The phenomenal rise in cable television, mergers and acquisitions at the networks, along with an unsettled union labor pool all laid the foundation for a paradigm shift in broadcast production and the labor force. A series of events occurred in rapid succession. In 1986, General Electric (GE) bought NBC through a purchase of RCA. GE reduced spending at NBC and then began to expand into cable television by its acquisition of shares in CNBC, Bravo, Arts and Entertainment channel, and others. At about the same time, ABC was purchased by Capital Cities for 3.5 billion dollars. Capital Cities then moved ABC into cable television by taking control of cable sports network ESPN. CBS was fighting off an expensive hostile takeover and the unions were threatening to strike. All three networks were anxious for a change!

This flurry of activity gave rise to the independent remote truck known as the OB van (outside broadcast van), which actively pursued work from the networks and cable broadcasters hungry for programming. This fleet of freelance television facilities was also happy to provide a talented freelance crew. Through all the mergers and acquisitions, network management changed and so did labor relations. The transition to a freelance labor began.

**Figure 1.3** NEP is a major supplier of television mobile vans. This unit is one of five OB vans used for NASCAR racing.

The freelance labor force got a boost in the 1980s with big demands from the networks and cable channels using a newly emerging industry of independent OB vans. The OBs provided top service, equipment and technicians and the industry grew exponentially. The final major shift in television production and labor relations occurred when Rupert Murdoch fashioned the FOX TV network from a bunch of independent stations around the United States. FOX shook up the stability of the big three when key talent, production and engineering left the "majors" and joined up, with hefty commitments and fresh capital.

**Figure 1.4** ESPN camera operator Jay Morrow fighting off wasp at an outdoor camera platform.

# The Quest for Sound

Over the years many audio mixers, or A1s, have been committed to pushing the envelope in audio production. In the early years, experience was lacking and audio personnel resorted to trial and error. Stories circulated about audio personnel placing a microphone in the golf hole on the green, in the baseball dugout, or in a condom underwater! I learned about the roar microphone from Dennis Finney at ESPN, when I had to climb 100 feet up the inside of the scoreboard at the Atlanta Motor Speedway to place a microphone. It had to be a cardioid microphone placed at the very top of the tower pointing up, so the microphone did not pick up sound reverberating inside the scoreboard. Television sound is a very demanding and sometimes frustrating job and should always be approached with a passion for adventure. The quest for sound begins with a love for sound!

The quest for sound is a journey. Television sound has reached new levels and will grow exponentially as high-definition television and surround sound become the norm. Surround sound offers unparalleled creativity as the move is made from *sound reproduction* to *sound design*. Sound reproduction is what is expected of a sound mix—for example, the sound of the bat crack is expected. Sound design often involves the unexpected and the subtle. For example, spatial placement is one of the new tools of sound design. Sound design is a quest for sound excellence and requires a passion and a drive along with a fine understanding for the art and science of sound (plus a good set of ears).

Television sound design is evolving as an integral part of the entertainment experience. For example, early radio and television coverage of sports had few natural effects and little atmosphere. Since those days, television sound has evolved into a medium that puts the viewer into the athletic experience. Consider the cameras and microphones in race cars, which have changed the viewing experience of motorsports.

## The Technology

The evolution of sound coverage and production parallels the improvements in electronics and sound technology, at a somewhat slower pace. The greatest advance in broadcasting was the development of small home speakers. When the consumer could finally hear the difference in sound quality on their television, they wanted more! Recent progress in the implementation of new technology in television sound has been pulled by consumer demands.

The implementation of surround sound has not been smooth. Many issues in monitoring, generating multiple mixes and upmixing two-channel audio sources are still being resolved every day through live television production. Surround was introduced in the early 1970s with an unsuccessful attempt at quad sound, but the problem for broadcasters was the delivery of the audio and video signal. Transmission of the television signal by satellites has a history in remote television, but analog satellite transmission has limited bandwidth and only two channels of audio. This not only hindered the progress of stereo but presented a unique challenge for any type of surround sound.

A major problem for the networks has been the existing infrastructure for moving audio and video around a facility. For every stereo signal path, only two wires were needed for sound, but with surround, six wires are needed for 5.1 sound. Control rooms had to be fitted with speaker monitoring, mixing capabilities and routing flexibility. Additionally, transmitting the audio signal and retransmitting the audio signal by affiliates had to be resolved. Early surround used encoding methods by Dolby and Circle Surround to process a surround-sound mix into a two-channel left-total and right-total enhanced stereo sound mix. The enhanced two-channel audio mix was compatible with any two-channel or stereo sound system, and a decoder was used to decode the two channels into a surround-sound mix similar to what was started with.

Encoded surround sound has presented mix challenges to the A1. Perceived loudness levels of announcers have been a significant issue for sound mixers because of improper speaker placement and calibration in OB vans and other small control rooms. Given all the problems that were being worked out, FOX produced its first season of NASCAR in stereo and worked on the issues associated with ramping up to surround sound in the background.

### Racing and Audio—Refining the Process

*Car racing has a long legacy of television coverage, with the Daytona and Indianapolis 500 garnering a large viewing audience. This large audience has allowed the network to invest heavily in these large-scale productions. In 1985 CBS Mobile Unit 8a and 8b covered the Daytona 500 with 12 cameras and a 32-Channel Ward Beck mixing desk. At the 1985 Daytona 500 Bob Seiderman used one microphone on a camera that was mounted on the wall at car level. This camera and the associated microphone became known as the speedshot and has been the source of great visuals and evolving microphone placement. A couple of years later, an additional microphone was added to capture the exhaust blast as the cars passed the camera. This combination of attack of the motor sound and the trailing roar of the car was an enhancement for the large speedways, because it provided a believable fill sound between camera cuts.*

**Figure 1.5** 1985 Daytona 500 CBS mobile units MU8a and MU8b plus a support trailer that carries cable and additional equipment and has a complete maintenance area.

**Figure 1.6** 2005 Daytona 500 with six mobile units plus generator.

**Figure 1.7** The compound diagram is enclosed in the production manual and the production manager will follow the layout of the compound consistently from venue to venue.

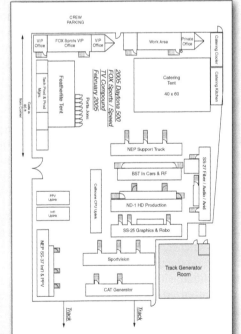

*Fifteen years later, FOX A1s Fred Aldous and Denis Ryan refined the sound of NASCAR and took it to another level. The same speedshot camera at Daytona now has three microphones—an omnidirectional approach microphone, a trailing exhaust microphone, plus a stereo AT825 mounted on the fence, away from the track for consistent left-right wider dimension. Surround sound certainly enhances the viewing and listening experience and FOX has led the industry in implementation and innovation of surround.*

*Innovation is the pleasure zone for audio. The speedshot microphone has been an evolving concept, and now with surround sound spatial placement is available. Fred Aldous places some of the trailing sound in the right surround channel, adding yet another dimension to the telecast.*

# The Workplace

Television studios in New York and Los Angeles produce sitcoms, game shows, talk shows and soap operas every day of the week and often in front of an audience. At the same time, location productions originate from all parts of the globe—ones as simple as a newscast and as complex as sports coverage in surround sound. The common denominator is that every single television production requires sound. Even though all types of productions are different in scope, methods are common and require the skills and talents of an audio practitioner when it comes to capturing and reproducing high-quality, engaging soundscapes.

Television sound usually includes a voice track that is captured live with headsets, lavalieres or hand microphones, depending on the type of production. On a soundstage, microphones are used on boom arms and follow the sound source around, while on a news set lavalieres are used because the sound source is usually stationary. All television production requires quality sound and all sound is prepared using microphones, mixing consoles, processing and playback equipment. The difference is in the mobility of the equipment. The single greatest difference between studio (inside productions) and remote television productions (outside productions) is the set-up and tear down.

Depending on the size of the shoot, a few hours to several days are needed to set up a television shoot. The tear down and packing of the equipment is commonly known as *strike*. Remote television production is usually sports or entertainment and generally occurs at a stadium or arena. Video engineering for entertainment has a familiar set-up, and as with any visual production, camera angles and lenses are chosen to deliver a desired look. Video and communications are effectively done in a television truck, but large-scale sound production often requires multiple specialized mixing positions.

# The Equipment

At the heart of any television production is the equipment and the housing for the technicians to operate it. Because of the high expense for broadcast equipment, there are very few fixed multicamera venues beyond the religious sector. That means that remote trucks must be brought in to shoot the production. However, much of the equipment used in studios and trucks is the same. Since the truck audio is more complicated due to its mobile nature, we will use a truck for our equipment discussion.

The television truck, or outside broadcast (OB) production van, comes in many sizes, varieties and specialties depending on the specific requirements of the production. A basic mobile unit is fitted with a video switcher, audio mix desk, intercoms, graphics, recording machines and cameras. For large television productions the television truck expands, almost doubling the size of the trailer. Specialty units are fitted with additional equipment that provides support to the main OB van. Television trucks are not optimized for sound monitoring and a specialized sound OB van will have proper space and speaker placement for sound monitoring.

The size, complexity and scope of television production have grown exponentially along with the related technology. As the scope has grown, so have the demands for more production space and expandable trucks. Extremely large and complex productions require specialized equipment and units that are specifically designed for that purpose. NASCAR coverage utilizes a great deal of "pop-up" telemetry visual inserts that require the specific capabilities of a unit that takes the audio and video after the team produces it and then inserts the graphics and sends the delayed video and audio back.

Depending on the type, length and location of the remote, an OB van may not be required and the equipment may be assembled and organized in what is termed "flight cases" or a "flight pack." Most broadcast equipment will fit into a 19-inch rack space and is easily assembled into portable cases for protection and shipping. Usually a systems engineer designs the signal flow of the production and specifies the equipment and components. The interconnect cabling has gotten easier with the use of fiber optics, and some existing systems may have wire looms of precut cable for recording devices and communications.

Whether using flight cases or a television truck, engineering set-up and systems checkout are always performed by the audio and video crew under the supervision of the tech manager, engineer in charge, video engineer and senior audio.

**Figure 1.8** Three-camera shoot on aircraft carrier.

Entertainment productions are generally musical variety, award shows, or concerts and tend to be music based. Music and concert productions require a sound truck with proper microphones and processing equipment to achieve an acceptable concert or studio sound that satisfies the artist. Often the rehearsals are recorded, so the artist can listen and tweak the mix for the final performance. Additionally shows like *American Idol* and the Grammys are in front of a live audience and the sound of the live PA and television are sometimes at odds. A monitor mix for the talent is critical to all television productions, especially a music production. The levels of quality and certainly the levels of expectation go with the Hollywood budgets.

The television truck becomes a temporary garrison of technicians and engineers and the area of operations around the truck is known as a television compound. The television compound will generally house one or more TV trucks, satellite uplink, office trailer(s), catering and portatoilets. There are similarities in all television or film productions particularly when it comes to logistics. The compound becomes a small city that performs a specific purpose and requires power, catering, telephones and security.

Safety is a serious consideration because of the large amounts of power, equipment and cabling that weaves a web around the television compound and often public spaces. Rigging of television gear often requires access to areas of extreme danger, such as high places and near hazardous machines. Television productions are often at venues with high noise levels and operators sometimes have to listen to communication at unsafe volume levels. Fiber-optics cable is used extensively to interconnect broadcast equipment and laser safety is a major consideration. (Never look directly into a fiber-optics cable to see if it is working!) RF emissions can be a problem around microwave and satellite equipment, which is prevalent around broadcasts. In addition, sports often continue during adverse weather and antenna stands and scaffolding for cameras can become a lightning-strike hazard.

## The Host Broadcaster

The international and national popularity of multiday entertainment and sports events has created demand for an uninterrupted television and audio signal of the entire event. The *international broadcaster* or *host broadcaster* produces core coverage of the event generating video and natural audio for all the "rights holders." The Olympics, Worldcup Football, and the Academy Awards are examples of events that utilize a host broadcaster. The rights-holding broadcasters overlay their own announcers and production elements and retransmit the final production in their home market.

The host broadcaster's role insures complete and unbiased coverage of the event, sport, athletes and entertainers and provides the core coverage so the rights holders do not have to duplicate resources. The host broadcaster minimizes the impact of seating loss by eliminating the need for multiple cameras at key positions. The host broadcaster also functions as an intermediary between the rights holders' requests from the event or venue.

The continuous coverage allows broadcasters to take commercial breaks at different times and a pay-per-view event can therefore be telecast without interruption. The host broadcaster coverage is recorded for archive and no action is missed.

## Audio Positions

A wide variety of job opportunities exists for audio technicians on soundstages, at broadcast facilities and from remote locations all over the world. The amount of sports and entertainment that is broadcast and recorded is staggering. One particular US network has three different production teams just for the regular football season. Hundreds of local and national television studios do some kind of original programming. Depending on the regularity of the production, the facility may have a full-time staff to cover the demands or use freelancers, or both.

The news and talk shows are produced and broadcast essentially live, although some of this programming is recorded and delayed according to time zones. While many of these programs tend to originate from the national networks, affiliate and independent stations in every major city across the world produce their own local news and interest shows.

Following are descriptions of the main audio positions in television sound production.

## Audio Mixer

The audio mixer (A1) needs a diverse set of skills that includes creative/artistic, technical and managerial. Creative skills are used in producing the end product, "the mix." Sound mixing requires the ability to judge the balance and quality of audio signals and requires the motor skills and talent to create a pleasing, effective soundscape. There is nothing technical about mixing—it is the artistic and creative aspect of sound. However, television audio also requires a technical mind capable of logical and methodical thought in order to set up, operate and maintain the large quantity of technical equipment. Technical skills are necessary.

Organizational skills are also essential to success for the senior audio, audio supervisor and AIC (audio in charge) on an OB van. Organizing and mapping the signal flow is critical to the success of a good mix. The audio mixer is responsible for managing the work load of the audio assistants.

Mixing skills are learned and practiced every time you sit behind a mixing desk. College basketball has been the proving ground for many top sound mixers and provides a good opportunity to practice the critical listening skills essential to constructing a good-sounding mix. Balancing the announcers with the dynamics of the crowd is a challenge with any production. Learning how to selectively listen to the producer and director is probably the most difficult aspect of television to teach, learn and train. "Ear-training" is learning how to listen critically and balance a sound mix, while selectively listening to the director and producer. The one day set-up and telecast common for college basketball is also good training for managing time and personnel. Additionally, basketball provides good practice on the set-up because the OB van is close to the venue, simplifying cable runs and connections in severe weather.

## Sub-Mixer

Large television productions will have one or more sub-mix positions. Award shows like the Grammys, Emmys, or Academy Awards will have a senior audio mixer managing and balancing the entire production, several music mixers, an audience mixer and maybe an RF microphone sub-mix. Some events have so many record and playback sources they may require an additional sub-mix position. NASCAR uses three mix positions and has a unique position that monitors the racing radios off air and feeds relevant conversations between the drivers and crews to the mix. This is an unusual position because it requires the mixer to act as the producer and decide what is relevant and fish for interesting conversations. This can be a gratifying role because it can make a big difference in the sound of a show without having to carry the responsibility of the senior audio.

## Music Mixer

The music mixer is a unique position because a knowledge of instruments, music and microphones is required. Microphone placement and equalization are often critical to the sonic characteristics of a music mix. Music production can require the reproduction of a mix that was created in a recording studio. Large music-award productions usually utilize a sound OB that is properly equipped for the live reproduction of music.

## Effects Mixer

The effects mixer is used in golf, motorsports, professional football and many large sporting events. Golf is a unique effects mix because the effects are mixed to air as well as sub-mixed and recorded for playback. A hole in golf will typically have between three and five microphones associated with it. These microphones are balanced and recorded off air for playback. Additionally, racing and golf are a very fast mix that requires intense concentration for a long period of time. There have been a number of studies about noise fatigue and, even though I am not a medical practitioner, I have experienced the drain of an afternoon in the truck!

**Figure 1.9** Effect mix position at NASCAR.

## Post-Production Sound

Post-production sound is common in film and edited video productions. Post production for a television sound track includes music, sound effects, additional voices or narration and fixing and previously recorded dialog or natural sound tracks.

## Studio Control Room

Productions like national news and weather usually use a "host set" where announcers present the stories between playback of video and sound. The sound mixer is responsible for the "live" sound from the announcers and mixing playback audio from record machines or video servers. Often music is played from nonsynchronous sources for feature stories or for going to commercial break. A challenge for this operator is the addition of reporters or announcers from a variety of locations. To prevent the announcers on location from hearing a delay of themselves, the host control room audio creates a mix of the program minus the location announcer and feeds that mix back to that particular location announcer. A separate audio mix-minus is created for each location announcer. For example, there could be two location announcers and a studio host. Location announcer number one will hear the host and announcer number two only. Location announcer number two will hear the host and location announcer number one only, and of course the host will hear everybody.

## The Audio Assistant

The audio assistants for sitcoms and game shows are the boom operator(s), playback effects specialist, and sometimes a dialog recordist. The audio assistant for a music production sets the microphones for the instruments, while the audio assistant for a sports production is responsible for helping set up microphones, run cables, and possibly operate a microphone during an event.

**Figure 1.10** *Saturday Night Live* production audio control room.

## Playback/Foley

Since the days of radio, the sound-effects specialist has been critical to the believability of a production. Before recording devices, the effects were created live by a "foley artist" using objects to recreate the sounds that matched the image on the screen. These objects could include things like coconuts for horse hooves. NBC's *Saturday Night Live* used NAB carts in the1970s, samplers in the 1980s, and now the entire effects library is played off a hard-drive playback. NAB carts are a continuous loop tape format that was used for music and sound effect playback. They used a tone to fast forward the tape loop from any place on the tape. If you had a four minute piece of music, it could take up to a minute to fast forward to the beginning of the piece. During a late night edit session in the OB van, this was often a source of frustration! Samplers are recording devices that are usually triggered by a keyboard. They can record a variety of sample sounds and the keyboard can instantly playback any sound. Samplers came from the music and recording industry and essentially were the first instantaneous playback device.

## Sound Recordist

In some productions the sound may be recorded on a separate record device and a recordist will set and monitor sound levels, listen to insure quality, check sync plus log and slate each take.

## RF Audio

The ever-increasing use of wireless microphones has created a niche position in the audio domain. Wireless microphones require constant attention because of problems with outside interference. Professional football productions will use a wireless audio technician on the field. There will usually be seven wireless microphones, one for each effects microphone, two for any on-camera talent, and one for the referee. The RF technician will also be responsible for the talent's wireless earpiece and any wireless communication circuits.

## Communications

Intercoms have turned into a specialty of their own. The majority of remote broadcasts use the intercom that is built into the television truck, but for large entertainment shows or events with multiple locations, additional intercom equipment and system specialists are desirable.

## Boom Operator

Hollywood has a tradition of using shotgun microphones on film and television sets, because the shotgun has a tight pickup pattern to minimize background ambiance. Most drama or situation comedy productions involve a lot of movement and require the specialized audio services of a boom operator. Operators need a combination of good motor skills and concentration.

The operator moves a mechanical arm and controls the movement and orientation of the shotgun microphone used to capture the dialog on a set. The boom arm is balanced on a pivot point that has 360 degrees of horizontal rotation and full vertical movement above and below the arm's plane. The boom pole collapses inside itself on a bed of rollers, giving the boom arm unlimited sweep of a set, only bound by the length of the arm and any lighting issues with shadows.

Small booms can be handheld by a boom operator for smaller productions. In Figure 1.11b, a boom operator is shown using a field mixer.

ENG/EFP (electronic news gathering and electronic field production) sometimes utilize field mixers. Portable cameras have gotten smaller, making it more difficult to monitor and adjust the audio controls. When there is multiple on-camera talent, a portable mixer is useful to control and monitor audio levels. Most quality field mixers will have a way to meter the audio and calibrate the mixer to the camera input levels. (See Figure 1.11a.)

**Figure 1.11  a)** A boom operator is shown working with a field mixer. **b)** ENG/EFP shoot field audio technician Chad Robertson.

## *Audio EIC (Engineer in Charge)*

Productions that are complex and have a nonregular crew increasingly have an audio engineer with the television truck. This has become necessary particularly with the use of digital mixing consoles that the audio mixer may not be familiar with. The audio EIC is responsible for setting up the mixing console, programming the communications, assisting with the patching and supervising the interconnect of any other television engineering and production vehicles or sites.

### *The Sound Designer*

The sound designer is the most senior audio role and is responsible for the planning and preparation of all the sound elements and audio concerns that will go into a production. During the preproduction period, a sound designer will meet with the show producer and director to insure that the production and engineering requirements are understood and that the proper equipment will be specified and budgeted for. The sound designer will deliver a wired and wireless microphone plan, evaluate the television facilities and recommend solutions to the production team.

All of the major broadcasters have senior level audio consultants that provide significant input for major events including crew, equipment, communications and the general sound design and philosophy. In many instances, key sound designers such as Fred Aldous at FOX, Ron Scalise at ESPN and Bob Dixon with NBC are shaping the sound of entire networks through their efforts. These people and many more talented audio technicians are basically writing the book, because every method and technology has changed and evolved in as few as four or five years.

## Breaking into the Business

Television sound has advanced far beyond what any single textbook or advanced television course can offer. There are plenty of books on microphones, but there is a definite lack of application information on microphone usage and placement for television. A few universities offer media degrees, but the problem is that the technology is advancing so fast that the universities cannot keep pace. To give the students hands-on experience, many universities have partnered with networks and broadcast facilities to offer internships.

When I began working in television, I was told that you first learn the workings of the outside of the truck, like the announce booth, microphone set-up and cabling. This is a practical and proven approach to learning television sound and how it is hooked up and brought back to the inside of the truck. However, learning the trade is not just learning how to plug things in, but also why! The knowledge of how microphones work should be gained from books and articles. The placement of microphones and what has worked best will be learned from other audio people you work with. Learn first, and then take any opportunity you get to prove your worth.

Entry-level audio positions exist as "audio utilities" for network television productions of college and professional sports. Utilities assist the audio technicians with cabling and equipment installation and during the event the audio utility will point a microphone at the desired sound source. Microphone pointers are common in American football. There is a lot of detail and work to be done in preparation for a broadcast, and this is a great opportunity to meet audio people and learn.

The audio assistant is the normal path to the mixing position in sports. The audio assistant learns the signal flow, equipment operation and production workflow by assisting the A1 and working with fellow audio assists. The audio assistant experience is invaluable to the A1 because it teaches troubleshooting skills and installation practices. I interviewed Dennis Ryan, who mixes NASCAR, and he said his experience as a field audio assistant taught him the sound of the tracks, how each track was different and how to approach miking a track. Some audio assistants have made a career of the announce booth and are valued by their A1s and the producers for keeping an orderly work area.

Television attracts experienced sound mixers from recording studios and PA companies. However, the only similarity between television and other sound mixing is the skill of balancing faders on a mixing desk. Music mixers usually successfully master television audio and find that their music experience contributes to their listening skills.

The Host Broadcast Training program originated with Manolo Romero and is maintained by Jim Owens for the Olympic host broadcaster in the host city. The host broadcaster organizes and conducts training programs, including a hands-on remote broadcast with television truck with the universities in the host city of an Olympics production.

Every professional should work on mentoring those who want to learn around them. Audio people are known for taking others under their wing and helping them advance their knowledge and career. Bob Seiderman was one of those that made a difference because of his exceptional talents and insight.

You have to stick it out! During the network's transition to a freelance labor force, the staff network technicians were often very unfriendly and made it clear that the freelancers were not welcome. The freelancers had to prove themselves, not only to their television peers and cohorts but also to management and production. Getting established can be a long and difficult journey. My advice is to take any job that will get you on the crew and take the opportunity to learn and practice. It will help if you are within travelling distance to a college town or a city with professional sports. Entry-level positions may not pay very much and are usually "local hire" or "daily hire," meaning the broadcasters are not expecting to pay for travel or provide accommodations for these positions.

Technology has advanced and most electronic equipment has changed from analog to digital. Fiber optics and advanced camera electronics now bring high-quality microphone signals with better technical specifications and fidelity to the digital mixing desk. Technical skills will help open doors because audio requires the application of technical intelligence and procedures. Basic electrical knowledge is essential for installation and maintenance of audio equipment such as microphones, power supplies, amplifiers and fiber-optics interfaces. Audio requires a significant amount of cable and connector maintenance, which mean soldering and metering are also useful skills. Finally, audio requires a lot of interconnecting cables, which are installed and derigged for each show. This means that a very basic skill for an audio technician is how to properly "over and under" coil cable!

Television is about problem solving. The more useful and knowledgeable you are, the more opportunities will make themselves available. Be ready and have your act together. Do not be the problem!

# 2 The Senior Audio in Charge

The responsibilities of the senior audio person (A1) will vary with the scope and size of the production. The A1 is generally responsible for the engineering set-up of the television truck, supervising the efforts of the audio assistant(s) (A2) and mixing the sound of the broadcast. The role of the A1 can be interpreted in many ways. The senior audio person needs good people and management skills, plus the ability to handle noncompromising deadline pressure. Additionally, it is essential to understand the television production process and be able to translate that knowledge into a successful production.

The A1 is the funnel for all audio-related problem solving. For example, when the directors headset is having a problem, the director will tell the A1 and the A1 will direct the resources to solve the problem. The senior audio has an obligation to keep the technical manager informed of the audio department's status. The audio supervisor is a technician, sound producer, maintenance supervisor and psychiatrist. This includes managing personality conflicts and strengths and weaknesses within the audio department. The A1 has one of the most diverse positions in television.

The A1 and the Director. *The director controls the flow of the telecast by organizing the camera sequence and is responsible for every video and audio element of the program that goes out over the air. While the producer may organize the content and sequence of the video and audio playback, the director calls the operation of the playback and puts the playback on air through the video switcher and audio mixer.*

The A1 and the Producer. *You have to listen to both the director and the producer to catch all the cues and the flow of the show. You can mix the best-sounding show of your life, but if you miss talent and tape cues, you will quickly be replaced.*

## Translating the Production Plan

The first task of the audio mixer is to translate the production needs into the signal flow of the television truck, including mixing-console capabilities and the inputs and outputs. The input capacity of the mixing console is consumed very quickly and some thought needs to go into how often an audio source is used and where to place it in the mix desk. This is a huge issue with video playback sources, because some programs may use as few as four video replay systems and a sports program may use as many as 20 replay machines. Add surround-sound sources and the mixer's capacity is consumed.

An entertainment production will have significantly different needs than a sports production, but there are basic audio threads in all television productions. The A1 is responsible for mixing the sound from the following sources:

1) Announcer(s) or dialog.

2) Audience or atmosphere.

3) Production-specific sound such as sound effects or live music if it's a variety or awards show.

4) Replay sources such as music and video with audio.

A basic sports production requires two announcers: a *play by play* and a *color* announcer. To support two announcers, three announcer headsets must be provided, which includes one spare. Two handheld microphones are needed for the announcers to use when seen on-air. This basic set-up is typical of basketball and baseball and requires five inputs into the mixing desk. A typical college and professional football game will include a sidelines announcer and an auto race may have as many as four announcers in the pits.

An entertainment show often includes announcers with wireless microphones or a podium microphone, and quite often has several live music performances. For large entertainment shows, the music is often sub-mixed in another sound truck and recorded to a multitrack recording format. In order to properly mix a large show, it is better to break down the basic sound elements into groups like music, announcers, sound effects, and playback sources. These subgroups are sub-mixed by a second sound mixer and then injected into the production show mix, which allows them to be mixed by the senior sound mixer. The senior audio directs and influences the sub-mixes, but the sub-mixes allow the senior audio mixer to concentrate on the overall sound mix and sonic quality of the show. The various mixes from the audio truck will be fed back to the television truck to be recorded or transmitted.

When laying out the mixing console, the mixer must organize the desk so that critical faders and knobs can be easily reached to make split-second adjustments and decisions at the direction of the director and producer. Analog mixing desks have become so wide that elements of a mix have to be combined and controlled by group mixing and master fader controls. For example, in NASCAR racing the sound-effect coverage for each of the six on-track cameras is captured using three microphones into four fader inputs (one microphone is stereo). The mix level of the four channels is controlled by one master fader but input level, balancing and processing are accomplished on four individual fader strips.

A sound mixer puts all critical mix sources and sound elements that need constant adjustment, along with group mixes and master control levels, within the span of an arm's reach. Digital mixing consoles solve the problem of mixing console size. Contemporary mixing consoles are designed so that, if desired, the operator does not have to move far away from the center of the console. Audio control functions are viewed on screens and the mixer can adjust signal levels, processing parameters and routing of the audio for any channel or mix group with a common set of function controllers. Accurate listening is achieved because the mixer does not have to move up and down the mixing console to adjust the channel strips.

The problem is that all digital mixing consoles are different and each one has a learning curve. See Chapter 3, "The Mixing Desk."

## The A1 as Engineer

The engineering role of the audio mixer is a matter of proper signal management and routing. This seems like a very sterile and unflattering definition, but this is the essence of one of audio's biggest roles.

Signal management is the beginning of the audio process and is something that needs to be planned and organized to insure a comfortable and quality work environment. Planning the signal flow, plus thoroughly documenting the signal flow and patching, will facilitate troubleshooting, maintenance and additional "oh by the ways" that may be needed. The OB van is designed to be completely flexible in the set-up and operation of the equipment. The patchbay contains all of the equipment's audio inputs and outputs. The patching process is essentially a matter of connecting or "patching" the outputs equipment, like microphones and playback devices, to the appropriate inputs of the mixing console and router. It is crucial that A1s have a well-organized and labeled workspace. Clear labeling of each input fader is necessary due to the large number of inputs available in today's consoles. On many digital consoles LEDs (light emitting diodes) are programmable for letters and numbers.

After the signal flow has been planned, the A1 will wire all the inputs and outputs of the equipment's audio sources to the appropriate destination. The audio inputs and outputs are centrally terminated in the audio control room's patchbay, which is the single location where all audio signals flow to and from. The audio patchbay is necessary to provide the flexibility for audio to be routed, split and terminated. Patchbays become very dense in large shows, making troubleshooting difficult. Cables can be labeled for quick and easy identification. Similar audio flow cables are grouped together in order to simplify troubleshooting. (See Figure 2.1.)

**Figure 2.1** Patchbays become very dense in large shows making troubleshooting difficult. Peteris Saltans like many sound mixers group and color code patch cord by flow or function.

The nature of patching and audio routing is changing to a central audio matrix scheme. Once the audio is input into the central audio matrix. it is accessible to the mixer, sub-mixer, recording device and for monitoring by programming the path.

## A1 Skills (Beyond the Obvious)

1. *Problemsolver:* An essential skill of the audio mix is the ability to troubleshoot and solve problems. This requires a logical thought process as well as patience. Often A1s know what a problem is, but it takes true skill and diplomacy to talk your audio assistant through a problem in order to solve it. Television is a constantly evolving situation. Up to the minute you go to air and after you are on the air, expect changes and additions. How the senior audio person deals with the pressure and reacts to changes will determine their career path. A1s have to learn to deal with the circumstances no matter what! In Figure 2.2, Mark Butler is pictured with two Shure 4 X 1 mixers when the digital mixing desk went out. The show must go on!

**Figure 2.2** When this digital mixing desk went dead, Mark Butler rigged two small Shure mixers together to get the job done.

2. *Ability to work under pressure:* You have to keep a cool head under the pressure. The audio person has to listen to the director and producer, mix a good-sounding show, possibly in multiple formats, and maybe even troubleshoot something while on the air. It is a tough gig! I have heard mixers throw up their arms in frustration and say "Mix it or fix it? I can only do one thing at a time!"

3. *Creativity:* This is not often associated with the skills of an A1, but an exciting, dynamic sound mix will draw viewers into the program. Interesting microphone placement and sound orientation differentiates one mix from another. The sound mixer has the ability to manipulate the soundscape and the creative challenge in television is to create movie-quality soundtracks live.

## Crew Communication

An often-frustrating part of the audio engineer's job is establishing communication between the engineers, operators and producers. Good communication is essential to the smooth execution of a live telecast. Audio communication generally is handled through intercoms that are established by audio personnel. Most intercoms are a point-to-point talk and listen, where a person talks directly to another person. The intercom systems are flexible and can also be designed as a talk channel, like a party line, where more than two people can communicate at one time.

A fairly common example of this is a dedicated intercom channel for video operators so the producer can talk to several videotape operators and orchestrate a sequence of video replays. A typical communication set-up would include separate intercom channels for the director, producer, engineering, cameras, graphics, video replay, lighting, audio, spotters, scoring and communication to the public address (PA) system. These individual channels are then routed to the appropriate crew members. For example, graphics should be able to hear the producer and the director, yet only talk to the producer. It is all programmable. See Chapter 5.

From here it only gets bigger, and some of the larger television trucks have systems with enough capacity for 196 personal intercom channels that includes programmable talk and listen. On an entertainment show, the communications is sometimes installed and maintained by a specialized group that may or may not fall under audio. See Chapter 5, "Communications."

## The Microphone Plan

Microphone selection and placement is a critical aspect of the sound design. Microphones are the foundation of the sound mix and proper microphone placement is critical for capturing the live sound elements of a television production. The microphone plan is the opportunity to examine the sound sources and plan the soundscape and sound design of the production. Microphone selection and placement is often based on experience and plans that have been used before and handed down by other sound mixers. The microphone plan is how the audio assistant learns good habits in microphone selection and placement.

Good ideas should not be abandoned. However, the microphone plan is not supposed to be a static document and should be improved as new technology and production techniques are possible. For many sporting events, the audio mixer is responsible for creating the microphone plan and may have precise instructions on microphone placement and brand. Many experienced sound mixers add their personal preferences and tweaks to a plan, making it their own.

For large international events, the microphone plan will be produced by a sound designer because the considerations for multivenue productions and stand-alone productions are different. Multivenue sporting events require a sound designer to produce a consistent sound between events and venues. Additionally, most events have controlling organizations or producers that supervise the look and safety of an event, including the placement of camera and microphones.

The microphone plan has to be implementable and safe. For example, hanging microphones from overhead fixtures is a preference in boxing because it provides strong on-axis sound downward toward the mat and rejects the noisy venue. In many situations, however, it is not practical to hang microphones because of location or time. The microphone plan has to be safe to rig and implement and also safe for athletes, entertainers or spectators to be around. The microphone plan provides clear instruction for the audio assistant as to precise microphone placement and selection.

The television truck generally travels with a minimum complement of microphones and other equipment and the sound design, including the microphone plan, helps to insure that the required equipment is accounted for and available.

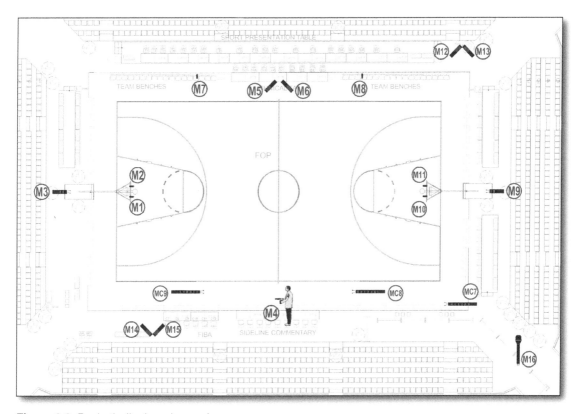

**Figure 2.3** Basketball microphone plan.

The basketball audio plan shown in Figure 2.3 is a stereo design using stereo and mono microphones. The foundation of this design is a microphone operator at center court to carry the ball thumps from side-to-side stereo coverage under the nets. There are two miniature lapels mounted near the net. This configuration is known as a spaced pair and provides a good stereo image for action under the net and shots onto the net hoop.

Stereo microphones on handheld cameras MC7 and MC9 give a dimensional left and right orientation to medium and close-up shots of the athlete.

Stereo-effects microphones M3 and M6 were substituted with stereo boundary microphones (Audio Technica AT849) mounted under the basket arm facing the court. Most mixers still use a microphone mounted on the basketball goal arm to cover the action under the net. This set-up provides a nice stereo spread and the microphone is out over the floor and closer to the action. Additionally, the stereo boundary microphone was very effective at picking up the coaches and spreading them left and right in the mix. This microphone also added a nice low-frequency sound to the mix and was used to feed the low-frequency effects (LFE) in a discrete surround mix.

Always use additional microphones specifically for atmosphere. There is a tendency to just rely on atmosphere spill into existing effects microphones.

Table 2.1 contains the list of microphones used in the basketball plan.

**Table 2.1**

|  | Mic Type | Location | Model # |
|---|---|---|---|
| M1 | Mini Lapel | In rubber under left basket | AT 899 |
| M2 | Mini Lapel | In rubber under left basket | AT 899 |
| M3 | Short Stereo Shotgun | Behind left basket | AT835ST |
| M4 | Super Shotgun | Operator | AT895 |
| M5 | Short Shotgun | On floor, far side, mid-court | AT4073 |
| M6 | Short Shotgun | On floor, far side, mid-court | AT4073 |
| M7 | Mini Lapel | Under coach's chair | AT830R |
| M8 | Mini Lapel | Under coach's chair | AT830R |
| M9 | Short Stereo Shotgun | Behind right basket | AT835ST |
| M10 | Mini Lapel | In rubber under right basket | AT899 |
| M11 | Mini Lapel | In rubber under right basket | AT899 |
| M12 | Stereo XY | Far side - crowd | AT4073 |
| M13 | Stereo XY | Far side - crowd | AT4073 |
| M14 | Stereo XY | Near side - crowd | AT4073 |
| M15 | Stereo XY | Near side - crowd | AT4073 |
| M16 | Hand Mic | Standup position | AT804L |
| MC7 | Long Stereo Shotgun | Camera C7 handheld | AT815ST |
| MC8 | Long Stereo Shotgun | Camera C8 handheld | AT815ST |
| MC9 | Long Stereo Shotgun | Camera C9 SSM | AT815ST |
| MC14 | Short Stereo Shotgun | Camera C14 | AT835ST |
|  | Stereo Boundary | Arm of goal post | AT849 |
|  | Stereo Boundary | Arm of goal post | AT849 |
|  | Stereo Boundary | Arm of goal post | AT849 |
|  | Stereo Boundary | Arm of goal post | AT849 |
|  | Stereo Boundary | Arm of goal post | AT849 |

## Supervising the Audio Crew

The senior audio person is also responsible for supervising and organizing the duties of the audio crew. A typical one- or two-day sporting event usually has an audio crew of three—the senior audio person and two assistants—while a large multitruck sporting event such as NASCAR racing or golf has a crew of up to 10 people: a senior audio supervisor, up to two additional mixers for effects, and as many as eight audio assistants to set up and maintain the equipment and operators. (See Figure 2.4.)

**Figure 2.4**  FOX's regular race crew at the 2005 Daytona 500.

An evening television sitcom generally has a crew of five. This crew would include a mixer, sound recordist, playback assistant and usually two boom operators.

### Facilities Check

Once the engineering set-up is complete, a comprehensive check of the equipment and each operators position is conducted. This is known as an engineering *fax* check (facilities check), in which all camera positions verify the quality of the communication channels and the program feed to hear the announcers and the video replay operators check communications and insure the proper input selection for recording devices. Every functional piece of equipment, communications and systems flow is tested one at a time. This fax check is the final proactive quality check before the broadcast. The audio mixer will then conduct a transmission check, which is a complete tone sequence and channel identification sequence that is sent through the transmission path to the master control room where the program is recorded and transmitted to affiliated stations.

# Mixing

The senior audio person is responsible for the supervision and implementation of all of the audio aspects of the engineering set-up of the television truck(s) and broadcast positions. Once this technical challenge is under control, then the audio responsibilities change to being a sound mixer whose mission is to create an esthetically pleasant soundscape that captures the essence and atmosphere of the event. As the sound mixer, the A1 acts as the audio producer and sound designer, which is a role of creative and quality judgment.

Sound mixing is the unique skill of combining and blending various audio elements to create an appropriate sound field that enhances the visuals. Additionally, live sound mixing requires quick reflexes, keen concentration and the ability to follow the action of the program and deliver each sound element of the production flawlessly. One of the greatest challenges for a sound mixer is to follow the commands of a director and producer and still balance a good-sounding broadcast.

Sound mixing is subjective and often difficult to describe and quantify in engineering terms. You can tune between networks and there will be differences in the tone and dynamics of each audio production. Generating a good-sounding mix takes experience and practice. A little experience will teach the beginning mixer how to optimize the mixing console for personal preferences, but time behind the mixing desk is the only way to tune the ear to the sound of the event and become accustomed to the constant chatter among crew members over the intercoms.

The fundamental job responsibility of the A1/Senior Audio is the blending and reproduction of the core sound elements of a production. While this may seem basic, sound mixing is not as black and white as it appears. Beyond basic reproduction is sound design and enhancement of the audio component or sub-mixed groups, which are sometimes referred to as "stems."

1) *Reproduction*—Certain sounds are absolutely expected in a mix. The quality sound of voices or the punch of boxing gloves is essential to the faithful reproduction of a sound mix.

2) *Sound design*—Sound design moves beyond basic sound reproduction to enhance the viewer's listening experience. This is accomplished with creative and nontraditional microphone placement, along with spatial imaging of the sound field.

3) *Sound supplementation*—Certain sounds are difficult, physically impossible or economically impractical to capture. This is when sound supplementation is a practical choice. The use of a sampler or the addition of sound effects to create a realistic ambiance would be examples of sound supplementation.

## Core Elements of a Sound Mix

Viewers and producers are paying more attention to a good-sounding production mix, which includes a good balance of the voice, atmosphere and production-specific sound. Capturing the natural sound of live television production is done with a combination of stationary microphones close to the desired sound source, plus microphones that are attached to the cameras and microphones that have operators. Proper atmosphere, sometimes referred to as ambience, and audience microphones are particularly critical to encapsulate the essence of the event.

In a television production sound mix, there are three basic layers of production sound:

1) Voice, announce or vocal.

2) Atmosphere (audience or ambiance).

3) Production specific effects. These include sport-specific sound, such as the sound of the kick of the ball for a football game, and a proper music mix for an entertainment act.

A good-sounding mix is often subjective but it is dependent on proper audio monitoring. Audio monitoring is reliant on the accurate reproduction of the electronic signal by the speakers and appropriate speaker placement. This is where the problems begin, because most mobile television production facilities must compromise on this set-up because of the lack of space.

## Sculpting the Sound

Another creative aspect of sound design is sculpting the sound using electronic improvements. Electronic processing has been used to enhance and/or control the dynamics, equalization and spatial orientation in a sound mix for over 60 years. Controlling the dynamic range of the recording process was essential because of the limitations inherent in early recording processes. Dynamics refers to the range between the softest to the loudest sound. When discussing dynamics within the context of sculpting the sound, compressors are used to smooth out the softest and loudest audio, ensuring that the sound is both heard and not distorted. Television programming is dependent on recording and fiber or satellite transmission for transporting and disseminating audio and video, which has a fixed dynamic range.

Equalization is used to adjust the tone or timbre of the sound source. If the sound is too "bass-y" (i.e., has too much low end) or has too much sibilance, the offending sounds/frequencies can be adjusted to minimize or eliminate the problem. Equalization is the principal tool in adjusting and balancing the tone of audio sources to match and sound appropriate. All mixing desks have an equalization section but it may be necessary to use additional gear outside of the mixing desk. See Chapter 3 for more details on mixing consoles.

Spatial processing with digital reverbs and digital delays is used to give a natural acoustic treatment to the sound image. Mono-to-stereo processors are often used to enhance a stereo image from mono microphones. Stereo processors are useful when creating a matrix-encoded surround mix and help prevent mono buildup in the mix. Additionally, digital delays are essential to bring audio in sync with delays in video sources. This processing can be accomplished by either an outboard piece of equipment or internally in the mixing desk.

A major goal for the Olympics has been a consistency of sound and tone between venues. At all surround-sound venues, the rear "atmosphere" mix was achieved from a pair of large-diaphragm studio microphones placed far left and far right.

### Equalization—Emphasizing Specific Frequencies

*Audio people, in the early days of AM radio, would emphasize specific frequencies so that instruments would cut through the other sounds. The Telecaster guitar was very popular with the recording industry because it had a bite to it that made it record well and was perfect for television shows like the* Buck Owens Variety Show, *where there were just a few microphones to capture the entire band.*

Electronic processing of the surround audio signal cannot be performed appropriately by using mono or stereo tools and requires a dedicated multichannel processor. Movies, where sound is generally oriented from front to back, usually treat the left and right front channels as a pair. The surround channels are also usually treated as a pair, with stronger compression in order to render the surround channels audible at lower listening levels.

In movies and surround audio programming, the center dialog and low-frequency effects (LFE) channels are compressed at a separate ratio from other channels. If the channels are not coupled together, the gain changes in the individual channels can shift the sound image. Stereo signal can shift as well and most compressor/limiter units and mixing consoles with dynamics have the ability to couple channels together.

A multichannel processor should be used for spatial placement of sound sources in surround space. Surround panning or advanced room simulation requires a close connection between the channels and using several mono processors will not work properly. You'll find more on sculpting the sound in Chapter 3.

## Juggling the Variety of Formats

One of the biggest difficulties for audio is the variety of sound formats that are broadcast: surround, stereo and mono are common for major events and productions. The variety of sound formats presents mixing and production issues for the A1 because the blending and combining of audio elements will sound and behave differently in each format. To accomplish a good-sounding mix, proper monitoring and metering will be required for each mix.

Even though the audio formats use the same audio sources, the common mistake is for the A1 to listen to and judge the quality of the sound production in surround, where the content is spread out between five or six speakers. A significant portion of the world population is still listening to mono television and radio. It is a challenge to get a good-sounding mix in which all the elements such as voice, music and effect come through with clarity in a single speaker.

Stereo mixing has become much easier today with affordable stereo microphones that attach to handheld cameras, mounted flat on any surface, or aimed at the desired sound source with operators. (See Chapter 7 on microphones.) The problem with capturing audio for sports productions is that you are always fighting to isolate specific sounds in very noisy environments. Stereo microphones, by their very nature, are broader and designed to capture a wider soundscape. A sound mixer can build a stereo sound image with combinations of properly spaced mono microphones and stereo microphones.

## Surround Sound

Surround sound is firmly entrenched in many parts of the world and most new home entertainment systems support multiple surround-sound decoders and are bundled with additional speakers. Sports programming has spurred the increase in numbers of households with surround sound and wide screen or HD television.

Additional audio channels and speakers give the sound mixer and sound designer space to create and orient a mix. Surround sound is the attempt to encircle the listener with audio to create a 360-degree envelope of sound that reproduces a believable soundscape and enhances the listening experience. Surround sound adds realism to the soundscape because it uses speakers in front of and behind the listener to give the impression that sound is coming from all directions just as it does naturally. Not only does surround add more speakers but also more channels of sound to create the illusion of spaciousness.

Basic surround sound is created using the traditional front left and right speakers plus the addition of a center speaker, a LFE speaker and a pair of rear-surround speakers. The LFE speaker should be considered an effects speaker because all the speakers can deliver full-frequency sound.

In a conventional two-channel stereo mix, the front two speakers deliver all aspects of the audio track: voice, events-specific sound such as the sport sound, atmosphere and ancillary music tracks. In a surround-sound mix, it is customary to use the center channel for the voice or announcer, front speakers for event sound, and atmosphere sounds to be added to the front channels and rear channels separately. As previously stated, the LFE should be used to enhance the low-frequency sounds such as explosions, punches in boxing, or even the kick of a ball in football. The LFE gives the additional punch needed sometimes to make sure the low frequencies are properly reproduced.

In a surround-sound mix there are no hard rules about the placement of music. During commercials I have heard the music tracks in all the channels. In a music video, I heard just the guitar track start in the rear channels and then move to the front channels. (I found this annoying and thought it did nothing to enhance the visuals.) For a television sound mixer, the job description includes the stipulation that the sound track should enhance the visuals and not distract from them. FOX Sports has taken it up a notch with Fred Aldous and Dennis Ryan. FOX Sports was the first network to telecast NASCAR in surround sound, incorporating creative microphone placement to enhance the surround experience.

## Surround Formats

Surround sound is delivered in distinctly different formats. *Discrete surround* is a multichannel format that is delivered in separate channels to the end viewers. This has been prevalent in Japan and is on the rise in Europe with satellite dish receivers. Dolby Digital/AC3 is a high-quality digital audio coding technology that enables 5.1-channel audio delivery to consumers, either via DTV, DVD or games. In DTV applications, it is used for the final transmission of discrete 5.1 audio to the home.

The mapping of the discrete audio signals in a 5.1 surround-sound format is defined as front left, front right, center, LFE, rear left, and rear right.

*Matrix surround* is a process that encodes the incoming five- or six-channel surround mix to a two-channel sound mix known as Lt / Rt – Left Total and Right Total. The two-channel method solves many of the delivery problems that occur with discrete surround when working with the massive two-channel infrastructure that exists in North America.

The two-channel mix can be delivered to viewers over cable, dish or terrestrial broadcasting. Those who listen in a normal stereo mode hear stereo and if the home viewer has a matrix decoder it will decode a surround-sound mix.

Dolby Pro Logic II® and Circle Surround are the major manufacturers of surround-sound matrix-encoding processing equipment. They then license the decoder software to home theater amplifier and speaker systems. See Figures 2.5 and 2.6.

**Figure 2.5** Dolby Pro Logic II surround encoders.

**Figure 2.6** Circle surround encoder.

## *Mixing Surround Sound*

Bob Dixon, sound designer for the Olympics at NBC, recommends the following mixing guidelines: "The surround channels should not draw attention away from the on-screen action." Constant, high-level use of the surround channel can be wearing and distracting to the listener and reduces the overall effectiveness. Dixon and Fred Aldous, a sound designer at FOX Sports, both approach surround sound with the same philosophy: keep it simple and use the best possible microphones.

The goal of the sound mixer is to create a soundstage in front of and behind the listener that is engaging and easy to enjoy. The front sound field should appear (image) from the left, center and right speakers to follow the image on the television screen. Localization, where audio draws the viewer to a specific focal point, is ideally not affected by the viewer's seating position.

To create a proper surround-sound mix it is necessary to have adequate mixer capacity and monitoring capabilities. The basic requirements for a 5.1-channel mixing/production is a mixing console with six discrete output buses for left, right, center, low-frequency effect, left surround and right surround (L, R, C, LFE, Ls, Rs) with panning between the five main channels and routing to the LFE channel. This set-up offers the greatest flexibility of sound placement in the surround field.

**Figure 2.7** The center speaker delivers a "hard" center, which gives good localization across the listening area.

The surround-sound field should sound natural and be used to convey a dimensional perspective and enhance the sense of depth and space. This is accomplished by using atmosphere sounds, music and crowd sounds. The surround audio mix begins as a stereo sound image of the event, including the venue atmosphere. Specific sound effects will be assigned to the front stereo speakers depending on the visual orientation and movement across the screen. Many events are shot from the perspective of the viewer, with most movement moving from left to right or right to left. FOX Sports' NASCAR coverage actually creates some "sound steering" from the front left to right rear speaker for added viewer impact.

The voice track has been a constant source of problems for the audio mixer. Finding and maintaining an appropriate blend is a challenge because live sound is dynamic and often unpredictable, particularly the voice track. Voice tracks are almost always placed in the front speakers so that the viewer's attention focuses on the screen. With a stereo mix, the sound elements are placed in both the left and right channel which creates a "soft" center that produces poor localization in the listening area. A "soft" center is caused by differences in arrival time across the listening area, causing a phase problem known as comb filtering effect. This can be heard by shifts in tone color, or a smearing of the sound image (indistinct sound). With a soft center, it can be difficult to manage a good blend of voice, audience and event.

When you use the center channel alone it delivers a "hard" center, which gives good localization across the listening area (Figure 2.7), but can easily be buried in the sound field. There is some consideration that all three front channels be used in some varied proportion for voice. Dolby Labs has devised a loudness level and metering that assist the audio mixer in establishing consistent levels of dialog.

One very effective way to achieve a stereo image with a visual reference is with stereo shotgun microphones on handheld cameras (Figure 2.8). Usually handheld cameras have smaller lenses and are generally closer to the action. This is perfect for a stereo shotgun because it does not have the reach of a conventional shotgun microphone due to the side microphone capsule. See Chapter 7 on microphones. Handheld cameras focus the audience's attention toward the screen.

**Figure 2.8** Handheld camera with stereo shotgun microphone.

Even though many events are presented from the left-to-right viewer perspective, handheld cameras, reverse angle and specialty cameras present challenges to audio personnel because of the changing visual perspectives. It can clearly be argued that changing sound perspective is correct to accurately match the video, but reversing perspectives can be confusing to the viewer and should be considered carefully.

Audio Technica makes several stereo shotgun microphones that can output true MS stereo and a XY stereo mode. MS, or mid-side, is more mono compatible and a truer phase-consistent method of driving stereo. This is important to know, because if the mixing console cannot mix MS, the XY can be used. Audio Technica also makes stereo boundary microphones that have been effective in recording music, sports and natural sound.

## Mixing the Crowd in Surround

Stereo and surround atmosphere and crowd should consist of a separate front crowd mix and a separate rear crowd mix. It is usually better to capture the audience and venue atmosphere sounds with spaced microphones at various locations around the venue. See Figure 2.9 which shows microphone placement at a large venue. This set-up avoids center channel build-up which occurs when the microphones detect too many similar sounds.

Dolby recommends this method of microphone placement when using Pro Logic II because it insures proper channel decoding to the front and surround channels and is a useful way to decorrelate the signal. Decorrelation essentially means keeping similar or like sounds from feeding the same channels, guaranteeing a better and more stable image in matrix formats.

When capturing the sounds of the crowd, microphones should be placed at various distances from the audience and positioned or panned to the left or right and front or back channels to set up the appropriate sound field. A variety of microphone positions will give the sound mixer several layers and textures of sound to blend for the appropriate sound scape. By adding closer shotgun microphones in front of a crowd, the mixer gives a closer intimate perspective to the sound; this is useful to give more definition to the front left and right sound field. Microphones placed further from the crowd tend to be diffused, making the crowd sound very large, like they are in a large stadium or arena.

Some sound mixers use shotgun microphones aimed across the field of play to keep a strong present sound with some distance. Caution must be taken when placing microphones too close to the crowd, which could pick up unwanted distinctive and individual sounds. The amount of surround-channel signal added determines how far back the listener is in relation to the front sounds.

**Figure 2.9** AT4050 Microphone at gymnastics for surround-channel ambiance.

**Figure 2.10** Japanese broadcaster NHK sets up a four-microphone array that feeds the surround channels for general crowd ambiance.

### Mixing the Audience at the Grammys

*In 2005, the producer for the Grammy Awards elevated the importance of the audience sound to the point that they included an audience sub-mix position. Klaus Landsburg, generated an audience/atmosphere mix in surround, stereo and mono. See Figure 2.11.*

*There were a total of 22 Neumann KM84 mics, four Sennheiser 416s, four AKG C547s across the front of the stage and two TLM 100s in the rear of the hall. Five Aphex 1788 8-channel mic preamps were remote-controlled via MADI, onstage.*

**Figure 2.11** Klaus Landsburg mixed the audience in 5.1 surround for the Grammy Awards.

The LFE channel is used when the sonic characteristics of the program will be enhanced by the low-frequency sound content. Low-frequency sound tends to be difficult to manage and meter, plus it generates high-energy sound levels that will quickly distort the audio signal.

A significant advantage of the LFE channel is that the television mix can operate at a normal level during the loudest part of the program and the LFE channel will create the desired impact without overloading the transmission path, speakers or amplifiers. This is useful during the coverage of fireworks, when the sound mixer often has to bring down the entire level of the audio mix to keep the explosions from overloading the signal path.

Wind noise can severely impact the use of the LFE. Most professional microphones have filters that roll off the bass in the microphone, but this also reduces the impact of the LFE. The LFE track is not included in the Dolby Pro Logic II signal.

Mixing is very subjective and impacted by personal and cultural tastes and preferences. I have found mixing styles and philosophies to be different around the world. The BBC has provided tennis coverage for the host broadcaster at the Summer Olympics for many years. At the 1996 Olympics, in my opinion the mix was a little reserved for championship tennis. I discussed this with the mixer from the BBC and he reminded me that this was what the BBC did at Wimbledon.

All sound mixes, especially surround sound, are subjective and vary with different broadcasters and even across international boundaries. As Bob Dixon previously said, "The rear channels should not distract the viewer's attention," but this is Bob's and NBC's opinion and is not a hard rule. My experience with the 2006 Winter Olympics found that the Japanese broadcasters preferred more "rear speaker" and LFE sound. I have found that mixing balance and tonal content is also affected by race and culture.

Currently, one of the greatest problems in surround-sound production is playback sources such as music and feature stories that are not produced for surround sound. Feature stories are normally full screen and full sound and break into the real-time program. When a playback source is produced in stereo or mono, the sound only appears in the front speakers and not in the rear surround or LFE. This is very distracting because the soundscape essentially collapses to the front speakers and then reappears in the rear speakers when the playback is finished. Commercials are produced in surround and exaggerate the deficiencies in surround-sound production

**Figure 2.12** A surround control room used by Japan's NHK.

Sound monitoring, and particularly surround-sound monitoring, is difficult in the confines of an OB van. The center speaker is critical for accurate dialog monitoring and essential in balancing dialog levels with the natural sound of the event (Figure 2.13). Even with proper monitoring, it is essential to compare the six-channel discrete surround mix with the stereo and mono sound mix. Most professional broadcast mixing desks will have multiple monitor selection controls to "A-B," or switch, between the surround, stereo and mono sound mixes.

Once again, the biggest mistake a sound mixer can make is to monitor the surround-sound mix and compromise the down mixes to stereo and mono. *Compatibility* between the mixes is a common term used, because the same input sounds are used in the surround, stereo and mono mixes. Signal *incompatibility* can occur due to phase problems. For example, when you combine two audio signals with similar frequency content that are 180 degrees out of phase, complete cancellation will occur! A partial cancellation can also happen which makes the audio sound like it is swirling around and is very unnatural.

Signal unsuitability can occur from improper monitoring. The sound mix may be perfectly acceptable in surround sound but may be missing a good balance in stereo or mono. This occurs because, in surround the sound, elements may be clearly audible because they are coming from multiple speaker sources. When you listen in stereo and mono, the sound is coming from fewer speaker/channel sources and is competing with other sound with similar frequency content. (See the sidebar on equalization.)

**Figure 2.13** Good positioning of center speaker for surround-sound monitoring and metering.

All matrix surround mixes must be listened to via a decoder to hear the effects of the matrix encoding process on the mix. Monitoring requires a properly calibrated speaker set-up for 5.1 reproduction and a surround-sound matrix encoder/decoder in the signal chain.

This is where the biggest differences between discrete 5.1-channel audio and matrix encoding systems become evident. A discrete sound mix is created when the various sound elements are assigned to the desired channels and remain intact within that channel assignment (spatial placement) through the distribution process to the end viewer. An encoded surround mix begins with the discrete channels and is processed into a two-channel derivative mix, which is transmitted to the user and then decoded back into the separate channels. This becomes difficult, because the derivative two-channel mix is also compatible with existing two-channel stereo playback as well as mono. Circle Surround and Dolby Labs deserve the credit for some amazing technology that processes a two-channel enhanced stereo mix to the end user and at that point it can be decoded into a surround mix. Remember, it is necessary to monitor the discrete sound mix, the Pro Logic II sound mix, as well as the stereo and mono mix to insure an appropriate and effective mix is delivered in that format.

While Dolby's Pro Logic II (PLII) and SRS aren't replacements for the discrete 5.1-channel mix, it is a step up from the Dolby surround matrix encoding system used for the last fifteen years. PLII offers advantages because premixed and pre-encoded elements can be combined to the left and right main mixing busses at the final stage. This is a significant benefit to producers and directors who heavily post-produce their coverage. For example, at the Athens Olympics, some venue sports were delivered to NBC and the other international broadcasters in an encoded matrix surround-sound format. The two channels of audio were easy to route around a television network facility through conventional two- or four-channel routing systems. PLII does not have to be decoded to edit and ultimately can be combined with other PLII elements.

The downside of any encoded matrix system is that decoders in viewer's homes can have problems decoding certain signals and may be improperly set up. Systems may experience a decoding problem in which channels may drop out or chasing occurs when the sound may not go to the correct channel. Other problems occur when the decoder decodes some of the stereo signal as random out-of-phase information and places it in the surround channel. Additionally, many home units have used controls that can completely change the mix from the original presentation.

## *Building a Mix*

Every time you add a microphone, whether mono or stereo, it will have an impact on the phase characteristics of the sound field and the sonics of a mix. Mono sound elements can be "widened" toward left or right with a short delay to offset the feed in one side. When used properly, stereo synthesizers can be useful for individual mono sources within a stereo and surround mix before encoding.

Do not rely on mono-to-stereo synthesizers in the final production or transmission path. Stereo processors and delay effects are particularly useful in sculpting a sound. Fred Aldous and many other excellent sound mixers use these tools and others to enhance the visual image by giving a greater sense of depth and realism or superrealism to the experience. Almost universally, processing will take place on certain elements of a soundtrack. For example, stereo synthesizers should not be used when dialog or vocals are part of the mono element unless this is the desired effect!

One of the big benefits of 5.1 surround audio is the ability to use a much greater dynamic range than in conventional stereo television. Everyone has been annoyed by the imbalance in sound between television programming and commercials. The sound mixer for the commercial only has to worry about a 30-second sound blast which can be remixed "ad nauseum." The live mixer has to deal with unpredictable volume levels and dynamics for typically over three or more hours at a time. Fatigue! Surround sound can be very engaging when the sound elements are creatively positioned in the mix.

Occasionally the peaks do approach 0dBFS (zero decibels full scale), the maximum peak level on a broadcast PPM meter. In a digital world, you cannot overload the digital pipeline because the audio turns to "hash," but do not kill the dynamic range of the program, particularly if delivering true 5.1.

There are great products like the TC 6000, which properly processes the surround signal and smoothly and accurately "peak stops" all channels of audio before overload. Compression has to work across pairs of audio. For example, the front stereo pair will interact between the channels because the sound does naturally in acoustics and is an integral part in order for the illusion of separation to occur. When the audience is on their feet in the front stretch of Daytona and when the band reaches a crescendo, it has to sound big!

## Up-Mixing Surround

A major issue for sound mixers in the surround world is using videotape or other sources that come to the production in a two-channel audio format or maybe even mono. Fred Aldous, of FOX Sports, stated that "It is very annoying for the production sound mix to collapse from a 360-degree sound field to just the front speakers." There is no perfect method to extract the music tracks and spread them out to a surround-sound mix without causing issues. The concept of "up-mixing" is on the front burner as the major broadcasters in the world are now all transmitting in some sort of surround. Up-mixing is the process of forcing a two-channel source into a 5.1-channel configuration. Pre-recorded audio tracks, videotape playback or any audio source produced as a two-channel stereo mix can be successfully up-mixed or "unwrapped." As a rule, the up-mixed process is better when being used for certain stereo elements within a 5.1 mix, rather than up-mixing a whole program.

FOX Sports uses a Dolby DP 564 Pro Logic decoder, which Fred Aldous feeds from two groups. The five-channel output of the decoder goes to the mixing desk and is blended into the production mix. The TC Electronics TC6000 processor is also used because it does a very nice job of expanding stereo sources to surround (Figure 2.14). I gained experience with this unit when NHK, the Japanese broadcaster, encouraged us to use the processor for the music of the opening and closing ceremonies at the Olympics. The prerecorded track for the ceremonies was mostly music but did include voice and vocal elements. During our up-mixing test, the voice and vocal tracks consistently were in the left, right and center channels.

**Figure 2.14** Technicians program a TC Electronics 6000 multichannel compressor/limiter to "unwrap" mode to up-convert the stereo music track to surround for the 2004 Olympics.

Up-mixing will be an issue for many years because of the use of archival footage. Music has generally been produced as a two-channel stereo mix.

## Down Mixing

Often television programs require multiple mix formats including surround, stereo and mono sound mixes. The mixes are often generated from the mix bussing of the mixing console. This method may require some compromise with individual fader levels because you are feeding multiple mixes from the same fader, but generally it can be effective.

Down mixing can be accomplished by combining or mixing the surround channels to a stereo mix. This can be accomplished with a variety of boxes including the TC Electronics 6000 and even the Dolby PLII encoder generates a stereo signal. Each processor has advantages and limitations. The TC Electronics platform is scalable and a single processor can adjust multiple audio processes from surround compressing to down mixing. NHK Japan has developed a system that is an adjustable mixer/combiner and they have done listening tests and have determined levels that work best for Japanese ears. The Dolby Pro Logic II encoder produces an enhanced analog Left Total/Right Total signal that is stereo compatible.

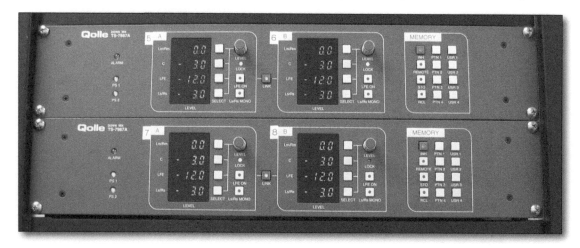

**Figure 2.15** The Japanese broadcaster NHK has developed a combination of down-mix levels of the 5.1 surround to stereo.

## Latency Issues

When audio and video is converted from analog to digital or up-converted from standard definition to high definition, there may be latency issues because of the signal conversion. Latency is a delay with the audio or video that is caused by the analog-to-digital conversion. For example, D-Cams are digital handheld cameras and the delay is about 20 frames, which is very noticeable. The sync problem can be compounded when the audio and video travel separate paths. This is typical of miniature cameras that go through frame synchronizers and are then up-converted. The frame synchronizer introduces a frame of delay and the up-conversion introduces a frame of delay, for up to two frames of delay. The audio may be a direct copper wire path with essentially no delay, so for perfect synchronization the audio will have to be delayed the two frames. Latency can accumulate to a point where lip-sync delay is noticeable (See Figure 2.16). (Diagram designed by Joost Davidson.)

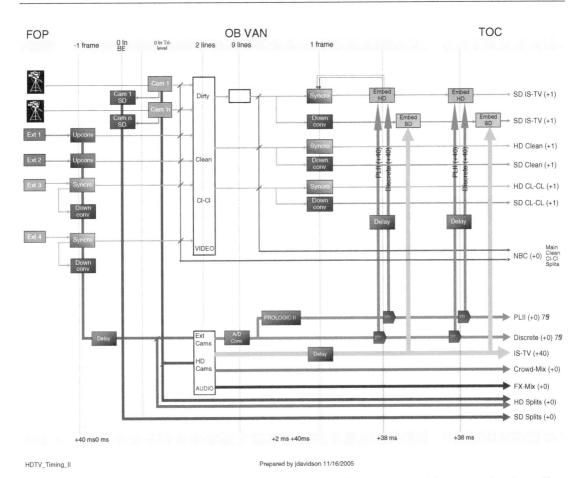

**Figure 2.16** Every video up-convert, down-convert, aspect ratio converter and frame synchronizer will delay the video and affect the synchronization of the audio and video.

## *Multiple Mix Positions*

During the 1990s Ron Scalise mixed all the NASCAR races that ESPN televised by himself and in a single truck! During the 2004 season, the NASCAR production used six television trucks: main production, in-car/RF, virtual graphics, audio submix, graphics and a separate television truck that does nonracing programming like pit-row shows. In addition, there is a generator truck, an office trailer and satellite trucks.

Many productions require multiple mixing with each mixing position. This requires a tremendous level of production focus and engineering execution. As entertainment and sports production has expanded domestically and internationally, the technical resources of a single television production truck (OBV) are maximized. The inherent beauty in the design of the OBV is its modularity. If a production needs more cameras, another OBV can be added with more cameras.

Very quickly, sports became playback intensive. Golf requires extensive recording to cover all of the players as well as to compress the action to a two- or three-hour broadcast. Audio for golf gave realism to the replay. As the system grew, the audio requirements swelled and golf production quickly began using two or three television trucks with a show mixer, a course effects mixer, and a mixer feeding replay machines.

The host broadcast mix for every Olympics is a sub-mix of the natural and sports sounds of a venue and event for the rights-holding broadcasters to overlay their local production. At the 2004 Summer Olympics, the host broadcaster generated high-definition video and surround-sound audio at thirteen sports. This introduced some new problems, ranging from sufficient inputs and outputs from the mixing desk to proper sound monitoring and metering in the television truck. Fortunately, the host broadcaster does not produce an announce track because most of the television trucks were set up for stereo and did not have a center speaker.

The Olympics host broadcaster produces surround-sound mix in a native five-channel format, front left and right, left surround and right surround and LFE, plus a Dolby Pro Logic II matrix output. Additionally, they produce a stereo and mono program mix and up to three separate stereo mixes such as a crowd mix without effects, effects mix without crowd, stereo handheld cameras, and several direct outputs as at each venue. HDTV venues were generally true HD video output, but included some 16 by 9 and up-converted video sources.

**Table 2.2**

| Sound Elements | Front Left | Front Right | Center | LFE | Rear Left | Rear Right |
|---|---|---|---|---|---|---|
| Announce | | | X | | | |
| Sports Specific Sound | X | X | | X* | X** | X** |
| Atmosphere | X | X | | | X*** | X*** |
| Camera microphones | X | X | | | X | X |
| Video playback | X | X | X | | X | X |
| Music and effects in 5.1 | X | X | X | X | X | X |

* Sound effects will use LFE.
** Rear effects may be used in sound design.
*** Separate and different atmosphere sound.

**Figure 2.17** Tim Davies of the BBC sub-mixes the track effects sound at the 2004 Games.

## Mix with a Live PA

Live events in front of an audience will usually have an extensive PA (public address) system. PA systems affect the television sound mix because the PA can be heard in any live microphones in the venue.

Cooperation with the PA mixer is critical for a successful television mix. Feedback from the house PA will end up in the television mix along with echo and delays of the original signal. The are several techniques that are used at most large award shows that include very fast gating and frequency band compression.

**Figure 2.18** PA mixer for the 2004 Olympic Ceremonies

# **3** The Mixing Desk

The mixing desk, or audio console, is the work center of the production, controlling the audio signal flow of the entire program. The audio mixer sets up the mixing desk to optimize the signal flow and monitoring points to properly manage the sound. The mixing desk is the nerve center of the operation and facilitates the inputs, outputs, signal level controls, routing and combining of the audio signals.

The outputs of the mixing console will feed a program show mix for transmission, usually surround and stereo, program or IFB mixes for the talent, the signal distribution router plus personal audio mixes for the director, producer and camera operators. Most television audio is stereo, but there is an increased demand for a dual feed both in surround and stereo. The stereo, or L-R selector routes the signal to the stereo output and is usually controlled by a single stereo master fader.

Virtually every mixing console has a similar flow and organization, with function controls at the top of the mixing console and the mix control faders at arms length in front of the operator. What is in the "signal chain" and the basic ergonomic layout depend on the manufacturer, but the basic functions and concepts of audio combining, blending, processing and routing are by definition what a mixing console is all about. The trend is undeniably moving to digital mixing consoles. Designs will continue to use more screens, menus and multipurpose control knobs and move away from the single function knob design.

**Figure 3.1** It is common to use two persons to mix large and complex sound productions.

Modularity and scalability has driven mixing-console design since the 1950s. However, a point of diminishing returns is reached with a "dedicated function" analog mixing desk. Mixing consoles can be organized into "dedicated-control functions" and "assignable-control functions." With a dedicated-control mixing desk, the controls for equalization, processing and routing are often linearly arranged in a straight row for quick manipulation.

In a dedicated function control mixing console, each knob and fader has a dedicated function, while a multipurpose control surface uses the same knob and screen for a variety of different control functions.

Assignable or multipurpose controls are often touch-sensitive knobs that change the screen to the parameters of the function selected. For example, if you touch an equalizer knob, the screen shows you the EQ curve and settings.

A plasma screen displays the control functions of the mixing console. Audio functions such as equalization, compression and imaging are displayed for each channel. Channel screens and display are programmable on some digital desks.

**Figure 3.2** Digital control surface with futuristic meter bridge.

The future of mixing consoles will be programmable work surfaces that network to a scalable audio matrix. The audio router matrix acts as an MME (master media engine) and directs all the inputs and outputs to the appropriate destination. The audio will be networked using large audio routers to provide the highest level of input, output and control capacity. No more patching record machines—the operator will be able to call up any audio source from the MME, adjust an appropriate blend of audio sources, and press record.

Open architecture is necessary in a software-driven environment because outboard production tools, digital archives and other audio-control platforms will need to communicate and exchange data with greater flexibility. Digital consoles will be required to interact in a plug-and-play fashion with all manner of computers, control devices and hardware. The audio work surface uses multifunction knobs to apply all the DSP (digital signal processing), and of course there will be long 100-mm faders to wrap your fingers around.

In a dynamic and innovative industry such as broadcasting, a sound engineer will be required to operate analog mixers, digitally controlled analog mixers, and digital mixers, as well as workstation control surfaces in a real-time environment.

No matter how complex and expensive the mixing console gets, there must be inputs for audio sources like microphone and playback devices. There must be outputs for routers, recorders and increasingly more surround sound and stereo mixes. However, the basic functions of a useful mixing console are controls for processing, routing and summing audio signals and everything else is design and features. The following section tells how we got there.

## History

The audio console has its roots in rotary attenuators and vacuum tubes. Before the early '60s, all audio consoles were custom built, hand wired and soldered together. Hard-soldered connections were tedious to troubleshoot and it was difficult to change out parts.

Bill Putnam owned and designed several audio studios in Los Angeles and is credited with many design functions and innovations that are still used today. Putnam designed a modular tube amplifier that was innovative because individual channels could be changed instantly with no down time. The plug-in chassis had a microphone preamplifier with a switchable line pad, low- and high-frequency equalization, and an echo send.

This approach standardized modular electronic circuits like the preamplifier, equalizer and volume control. It also organized the audio functions in a linear block in order to facilitate the operation and maintenance of the mixing console. With a common power supply and audio routing, this became the foundation for the modern day in-line console.

In 1961 Rupert Neve assembled two tube consoles for Recorded Sound in London. The requirements were for a console with high sonic quality and which was extremely reliable and transportable. Neve's approach to modular in-line circuitry was similar to Putnam's, but Neve pioneered a product that delivered superior sonics and portability. ESPN used Neve consoles in all five of their original OBs and Doug Dodson, original Engineer in charge of OB-140, does not remember them ever having a failure.

Solid State Logic (SSL) had one of the first dual fader in-line consoles and it was used by NBC and Turner Television because it packed many fader controls into a small space.

Calrec was formed in 1956, originally designing amplifiers and microphones, but by 1970 they began making broadcast-specific mixing consoles for the BBC. Calrec showed a digitally controlled analog console to the BBC in 1984.

Early television broadcast trucks (OBs) were designed by network engineers who adapted basic audio and video studio equipment to the remote broadcast. The design of professional audio mixing consoles had standardized into a modular approach that used in-line electronics modules mounted to a scalable frame using multipin connectors. Each electronics module was individually powered from redundant power supplies that minimized and localized any failure. This appealed to the broadcasters, who could anticipate maintenance requirements and carry plenty of spare parts along with an engineer to do maintenance.

Mixing consoles have been traditionally designed for recording studios or public address systems, and more recently specifically for broadcasting and theater. The recording industry prompted the rise in independent studios in the early 1970s, which drove the console business. The race was on to pack as much as possible into the next generation of mixing consoles.

Recording consoles found their way into OB vans along with some of the other equipment used today. Tape returns, multitrack assignments and foldback are oriented to the recording studio and most recording consoles were designed to have a "split" or an "in-line" mixing and monitoring design for multitrack audio tape machines. A split design has an input section for microphones and a separate line input section to monitor line-level tape machines. Usually in a split-mix design, there were completely separate mix paths for the microphones and tape machines, which made some operations more difficult or impossible. A split fader or dedicated input design was used by Soundcraft, AMEK and others.

**Figure 3.3** PA manufacturers like Yamaha earned reputations for reliability with concert touring and became popular in broadcasting because they offered many inputs and lots of mixable and prefader outputs.

Until the late '80s, CBS used Ward Beck consoles in their OBs. Bob Seiderman mixed the Daytona 500 and many Super Bowls on this basic 48-channel, 12-group Ward Beck mixing console (Figure 3.4).

**Figure 3.4** Ward Beck Mixing Desk in CBS OBV 1986.

The biggest advantage of a conventional analog console is ease of use. Generally, all channel strips are identical with dedicated function knobs and moving a signal through an analog desk is fairly logical and straightforward. The problem is that a dedicated function-control approach will reach an impractical size in a hurry! Every time you want to add another channel of audio, another entire channel strip has to be added. This not only increases the size and weight of the mixing console but, as the console becomes longer, proper monitoring becomes an issue and the sound mixer will quickly wear out from moving back and forth behind the console.

Increasing console capacity while maintaining a usable frame size for OBs has been a significant challenge. ABC Sports installed an SSL 4000 mixing desk in their OB van that debuted at the 1984 Olympics in Los Angeles. This was a real turning point in the size of television productions and emphasized the need for a higher level of mixing desk.

The SSL 4000 was one of the first mixing consoles to incorporate a second fader in the channel path, essentially doubling the capacity of a mixing desk (Figure 3.5). With the addition of 48 line-level inputs, the SSL offered a great solution to the broadcasters with flexibility, modularity, reliability. It was a big step up from previous consoles in quality.

Dual fader in-line consoles are designed so that each channel strip has both a variable level input signal path with input gain from microphone to line level, plus an additional line-level tape monitor signal path. These are entirely separate signal paths with two faders in the channel strip, usually a large fader for the input path and above it a small fader for the tape monitor path.

Each channel strip was capable of switching between each fader path, allowing equalization and dynamics to be used by either channel. (Note the analog SSL could switch between each fader path, but could only be used by one or the other.) Of course, each fader had separate routing and signal flow paths.

**Figure 3.5** SSL 4000 (Solid State Logic) mixing console in ABC Sports OB van.

Mixing consoles have benefited from better design and improved electrical components and each new generation of mixing desk does more and is smaller. Once a signal flow matrix was established, the size and capacity of mixing consoles grew exponentially to satisfy the increasing demands of large-scale studio and television productions. However, the size and weight of mixing consoles grew substantially and quickly reached a point of diminishing returns.

The television production truck has limited space and designers are always pushing to get more inputs into smaller surfaces. The audio requirements for an entertainment or sports production might include multiple announcers and announce positions, plus production-specific sounds and atmosphere along with a myriad of playback sources. Audio playback sources and the number of video recorders have risen exponentially, with some digital recorders capable of playing eight channels of audio. Audio for surround consists of six channels for discrete and two for stereo or matrixed surround.

Often multiple-mix positions are used to accomplish high-capacity productions, but it would take a radical departure from analog console design to meet the next level of mixing demands.

In order to reduce the size and cost of a mixing console, it would be necessary to reduce the dimensions and number of electrical components or eliminate redundancy in the design. Audio signal blocks like equalizers were identically repeated on every fader channel, which required tremendous space. Digital circuits and the advent of microchip DSPs (digital signal processors) provided increased control and flexibility in processing audio signals and eliminated the need for redundant functions like equalizers in every channel.

Digital mixing consoles have solved the problem of incremental increases in faders that resulted in exponential increases in mixing console size. Digital mixing consoles eliminate the need to have the audio inputs, processing and control in a single frame.

**Athens Greece—Opening Ceremonies 2004**

**Figure 3.6** This Soundtrax Mixing console was used for the 2004 and 2006 opening and closing ceremonies. The desk has dual in-line large faders in the flat surface closest to the mixer. The Soundtrax mixing console has 96 mix faders, six function screens plus analog and digital inputs and outputs. Digital consoles have the huge advantage of instant recall, which completely eliminated the problems of musical acts with different volume and EQ settings.

*Assignable digital control strips makes it very easy to operate large numbers of channels that can be packaged small enough for the OB van, which makes more room for the A1 and proper speaker set-up.*

## Basics of Mixing Consoles

Analog mixing consoles have a signal path that flows linearly through a succession of amplifiers, processors and summers and will always have dedicated control functions to adjust the audio signal (Figure 3.7). In an analog mixing desk, the audio signal is always analog. Digital mixing desks convert the analog audio signal to a digital fingerprint of binary numbers that is easily manipulated. Digital mixing desks benefit from the ability to completely randomly assign functions, controls and signal path.

All mixing desks, or consoles, have: inputs for microphones and/or line-level sources, plus some form of tone control or equalizer, output routing and a mix fader. Professional features and options might include insert points, auxiliary sends, dual faders, dynamics controls, panorama capabilities, Solo/PFL and various output configurations. Stereo microphone channels will have two phantom power sources and possibly MS stereo decoding.

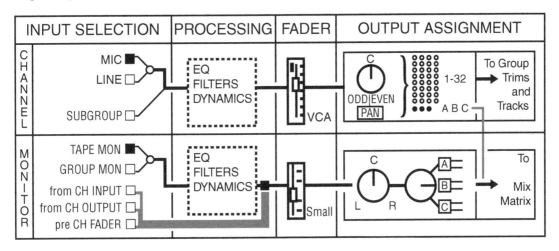

**Figure 3.7** This is a block diagram of an SSL 4000, which is typical of a dedicated function, dual input, in-line analog mixing console.

The SSL 4000 (Solid State Logic 4000) was innovative in design features, such as dual input faders per channel strip. The microphone input section amplifies the low-level microphone signal to a line level so the electrical circuits can manipulate the signal and route it to other sections for summing or output. The second fader, also known as the top or small fader, is a monitor input for line input sources and does not have enough gain or amplification for microphones. This dual-input logic has been incorporated into many design philosophies.

## Console Inputs

A basic function of a mixing console is to get a good clean microphone signal into the mix desk. The microphone input will supply 48 volts of phantom power to the microphone and the power should be switchable on/off. Additionally, the input section will have a phase button that inverts the signal and a pad button that attenuates the signal usually by 20 dB. (See Figure 3.8.)

**Figure 3.8** Lawo MC90 uses a dedicated input control section per channel strip. The six function buttons are phantom power, line input, pad, microphone input, filter and phase. There is a single rotary knob for level adjustments.

Audio signals are low levels of AC (alternating current) voltage generated by microphones and playback devices. Microphones deliver a normal voltage/signal as low as 1 mV and up to around 250 mV. Low-level microphone signals require a high-quality, low-noise amplification circuit to strengthen the low microphone levels up to line level, or around 1 volt. An amplifier, such as a microphone preamplifier, is an electronic circuit that produces gain/amplification by increasing the voltage of the signal being processed.

Microphones will have a variety of output levels depending on design, construction and even age. A variable gain control on the input of the mix module is essential for proper gain structure, and some consoles use a single input with variable gain from a low microphone level to a line level (Figure 3.9).

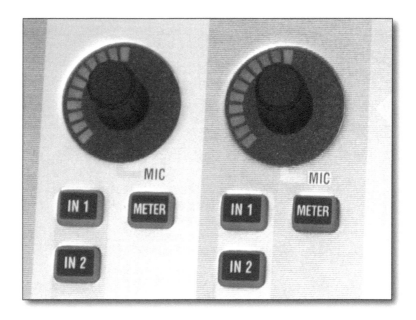

**Figure 3.9** Simplified input control with input select and a rotary control with LED level indicators.

Line inputs generally accept signals of around 100 mV to 1 V and generally provide a minimum of around 30 dB of amplification. (Electrical circuits are measured in volts of electricity. Audio signals involve such small increments of electrical voltage that they must be measured in increments of a volt, known as millivolts.) Low-level audio signals must be amplified to line level to be manipulated by the internal line-level circuitry of the mixing desk and other outboard processing gear. Additionally, analog from microphones must be digitized to be processed in a digital mixing console.

Analog mixing desks have the microphone preamplifiers built into their architecture. With a digital mixing desk, electrical components, amplifiers and display functions can be rack mounted, leaving a smaller control surface in the audio control room. A digital console can use analog, optical or digital interfaces, which permits the designer to put the microphone preamplifier, phantom power and analog-to-digital converter in a remote "stage box." Once an analog audio signal has been converted to a digital audio signal, processing and routing becomes easier and more efficient.

### Digital Mults

*For the audio crew a digital/optical stage box as close to the microphone sources as possible minimizes interference for electromagnetic sources, speeds up the set-up and reduces the amount of heavy copper that has to be carried on a remote. This remote box has a fiber system that provides eight channels of audio and intercom on one fiber and camera on another. This fiber system is an analog input at the source and an analog output in the broadcast compound where the audio is distributed to the various destinations in analog.*

**Figure 3.10** Field connection box with fiber-optics encoders for audio and video.

*Many digital mixing desks have a direct digital input into the mixing desk and can be as simple as a fiber or digital wire connection direct from the stage box. Some digital consoles use a routing matrix to input all the audio sources directly and then the digital audio signal can be assigned or routed to the appropriate destination on a screen.*

*There are several "open" standards for digital interconnect that console manufacturers have adopted digital interconnectivity with the AES10, MADI, and AES31 communications protocol. MADI provides a full digital link on up to 100 m of coax or 50 m of optical fiber to remotely place the interface box.*

## Measuring the Signal

The term decibel (dB) is a measurement used for sound gain or amplitude, but it is also widely used to measure power, voltages, sound pressure level and more. Because the decibel is used to calculate a variety of sound measurements, its interpretation is complicated by different measuring scales and references. When a sound is generated by natural means or by some amplification, the intensity of that sound is measured in decibels of *sound pressure level* or SPL. Measuring sound pressure is useful in microphone selection, but is not a measurement used in electrical audio equipment.

The output of electrical devices such as microphones and mixing consoles is an electrical image of sound and the intensity of the electrical image is measured in voltages and expressed in decibels. The decibel is a measurement of the ratio between a level being measured and a reference level.

Broadcast engineers and technicians will encounter several decibel measurements for audio equipment. As digital equipment replaces older analog gear, the decibel full scale (dBFS) will become the common measurement and standard. However, till this day occurs the audio engineer is required to interface a variety of analog and digital equipment, which measure and meter audio with different scales.

**Figure 3.11** Analog Volume Unit (VU) meters and Plasma meters give the sound mixer a quick and accurate view of changing volume levels.

Analog audio equipment measures the output signal in voltages and derives a decibel measurement by comparing the output voltage to a reference voltage. Audio signals are usually measured as RMS values because they are similar, but not the same as average audio levels. RMS (root mean square) is obtained by squaring all the instantaneous voltages, averaging the squared values and taking the square root of that number. When we want to evaluate the loudness of a signal, RMS corresponds closely to the sensitivity to our ears.

High-impedance audio equipment is sensitive to voltage and generally uses dBV to measure the ratio of audio signal level compared to 1 volt RMS (0 dBV = 1 V). There is no reference to impedance because this measurement is generally used with high-impedance equipment, which is sensitive to voltage.

Early broadcast and studio equipment was low impedance and sensitive to power and dBm was used to represent the power level compared to 1 milliwatt (mW). This is a level compared to 0.775 volts RMS across a 600-ohm load impedance. This is a measurement of power, not a measurement of voltage. Most analog mixing consoles specified nominal operating levels at +4 dBm for the US and +8 dBm for Europe.

Since a majority of audio equipment will operate at a wide range of impedances, an "unloaded" measurement was devised; dBu represents the level compared to 0.775 volts RMS (u = unloaded). This measurement is used with an unloaded or open circuit source, which is typical of the insignificant load from high-impedance audio equipment.

Remember this! A low-impedance line output can generally be connected to higher impedance inputs without much signal loss, but a high-impedance output connected to a low-impedance input may distort and overload the signal.

A table showing relevant conversions and more detailed definitions follow.

**Table 3.1** dBV to Voltage RMS to dBU Conversions

| +12.8 dBV = 4.4 volts RMS = +15.0 dBu or dBm |
|---|
| +6.0 dBV = 2.0 volts RMS = +8.2 dBu or dBm |
| +4.0 dBV = 1.6 volts RMS = +6.2 dBu or dBm |
| +1.78 dBV = 1.23 volts RMS = +4.0 dBu or dBm |
| 0.0 dBV = 1.00 volts RMS = +2.2 dBu or dBm |
| −2.2 dBV = 0.775 volts RMS = 0.0 dBu or dBm |
| −10.0 dBV = 0.316 volts RMS = −7.8 dBu or dBm |
| −20.0 dBV = 0.100 volts RMS = −17.8 dBu or dBm |

*dBFS*—Digital full-scale is used to measure amplitude of an audio signal compared to the maximum signal that a device can handle before clipping. This is a peak measurement because, once the peak is reached, circuit saturation will occur along with digital distortion and unusable audio. In a digital circuit, 0 dBFS is equal to the maximum level the processor is capable of expressing. For example, 16-bit audio has a dynamic range of 96.33 dB and when the audio circuit is reaching a maximum input level of 96.33 dB, it has reached 0 dBFS. This designates the highest possible level, and all other measurements expressed in terms of dBFS will always be less than 0 dB.

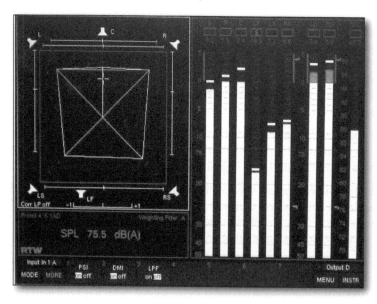

**Figure 3.12** Surround-sound metering with bar and dimensional displays.

This is why, on digital gear using VU meters (where 0 dBVU means 0 dBFS), the "0" is at the top of the scale and the meter can never read higher than that.

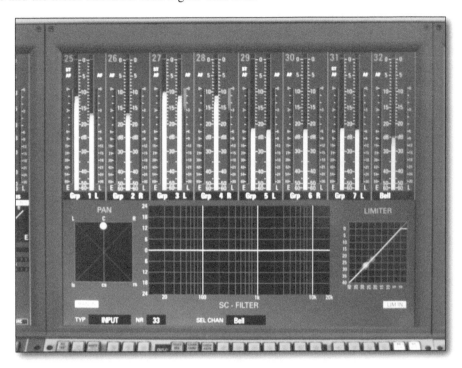

**Figure 3.13** dBFS measurements. 0 dBFS is the absolute peak where the audio will become unusable. Set-up for 0 dBFS is normally at –18 dBFS to give 18 dB of head room.

*dBSPL*—A measure of sound-pressure level with 0 dB SPL defined as the threshold of hearing. The decibel is a good measurement of sound levels because the ear is capable of detecting a very large range of sound-pressure levels and the human perception of sound intensity is roughly logarithmic. The normal range of human hearing extends from about 0 dB to about 120 dB, where 0 dB is the threshold of hearing in healthy, undamaged human ears. The ratio between the threshold of hearing and the threshold of pain is over one million to one. Logarithmic scale is a simple way to represent very small numbers with a lot of digits.

The human ear has a sensitivity range of about 120 dB and a 3-dB increase in sound intensity is barely noticeable, while a 5-dB change in level is easily perceived and a 10-dB increase in the level of a continuous noise is perceived to double the loudness.

# Metering the Signals

Television sound should be dynamic and similar to real life. Unfortunately, audio equipment does not have the same ability to process vast dynamic ranges of sound as the human ear.

Measuring and metering the audio signal is critical to avoid saturation of the audio circuits and distortion and is the reason why all professional audio equipment and mixing consoles use some visual references or "meter" to visually monitor audio.

Plasma screens and mechanical movement type meters are commonly used to view the relative audio levels, but the measurement scale and measurement source must be understood to be relevant. Analog mixing desks and digital mixing desks will both have plasma screens but the reference level and measurement scales are completely different.

**Figure 3.14** Plasma and analog VU meters are found side by side in this SSL200 series console. Long PPM meters at top left and Ballistic type VU meters at top right give an engineer two different visual references.

The audio VU meter is a familiar-looking device that measures dBV or dBu and is a good visual reference for loudness. Meters measuring the output level of recording or audio gear are usually measuring AC RMS voltages. Most analog audio circuits are designed with a wide range of operating capabilities to accommodate dynamic audio signals. A measurement of 0 on an analog mixing desk is usually 18 dB below the circuit saturation point.

## Peak Program Meter (PPM)

This type of meter accurately measures the peak characteristics of the audio signal. The traditional VU meter does not show peaks accurately but corresponds quite well to the subjective loudness of a signal. The design and movement of the meter is what is known as the ballistic characteristics of the VU meter and analog meters tend to be slow compared to a PPM type of meter.

There are a number of metering functions that audio personnel must be familiar with when working with an audio console. The most critical measurement is the input audio level to insure good audio gain structure through the signal path. PPM and VU meters are the predominant metering device, but plasma screens are replacing traditional meters because they provide quick snapshots of user-determined audio information (Figure 3.15).

**Figure 3.15** Plasma screens offer a lot of information in a concise, easy-to-read view.

Plasma PPM meters offer fast, accurate signal monitoring in a dense format. The mixer can easily scan the 24 meters for input levels. Plasma meters offer features that cannot be provided by ballistic-type VU meters.

*Display peak:* shows the highest level reached as a single segment at the top of the meter column.

*Store Peak:* retains the highest level reading reached in the channel, until cleared by the "clear peak" button.

Digital consoles allow the users the flexibility to arrange the order of processing (i.e., equalization, dynamics or delay) of the signal flow and permit metering and monitoring of the audio at virtually every stage. For example, it is common to want to meter input audio levels to optimize gain before the preamplifier or analog-to-digital conversion. Overloading the input stages of the mixing desk will cause analog and digital distortion, which essentially renders the audio unusable.

### Gain Structure and Headroom

*There are many stages in a mixing console where the signal level can be altered. The first and most critical stage is the input section and the microphone preamplifier, where proper gain structure is the difference between distortion and excessive noise. The electrical levels entering a mixing console vary tremendously depending on the sound source and the method of inputting the sound. Professional mixing consoles are designed to deal with extreme microphone output voltages across the input circuits. Proper input gain is critical in digital consoles because, once you have exceeded 0 dBFS, the audio signal becomes unusable.*

*The useful range of a mixing console can be extended by inserting a pad to attenuate the signal before the preamplifier. If the microphone is overloading the input of the microphone preamplifier, the audio signal will be distorted throughout the signal path.*

*Additional gain comes from send controls, summing amplifiers and even equalizer circuits, which are "active circuits" and electrically alter the signal.*

*Motorsports audio is so dynamic that audio level adjustments are made before, during and after a race. The sound mixers have computer control over the gain electronics of cameras, but often pads have to be inserted before the camera microphone preamplifier to avoid overloading this input stage.*

*There are never enough mix inputs in the remote truck and often a second mixer must be used to sub-mix parts of the audio. Gain structure is critical when cascading sound mixes through mixing consoles. To insure proper gain structure through the signal chain, test tones are used.*

## Shaping the Signal

The most common processing of the audio signal is tone adjustment or equalization. Equalization is usually accomplished with a parametric equalizer, where there are usually three parameters that the A1 will be able to control: frequency, Q and level adjust. (See Figure 3.17.)

*Parametric equalizers* are equalizers that are adjustable around a specific frequency where there can be a boost or cut in tone. Adjusting the *frequency* changes the tone characteristics you want to accentuate or cut. A parametric equalizer can precisely tune to a desired frequency and adjust in a very narrow frequency spike. If you widen the peak/center of the frequency, a bell-shaped curve develops on both sides of the peak of the frequency. The frequencies under the bell are affected by the equalization process. The width of the bell curve or the range of frequencies affected is an operator control function and is commonly referred to as Q. Q is a measure of the bandwidth of the bell-shaped curve.

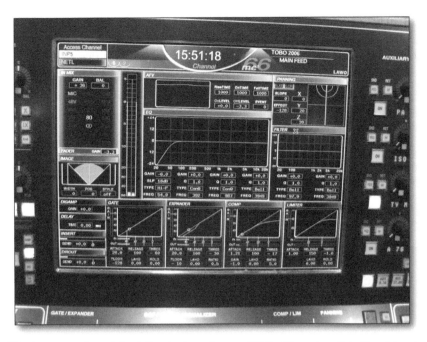

**Figure 3.16** This is a snapshot of the left net microphone—NETL. There is a 6.0 dB cut beginning at 100 Hz and the gate, compressor and limiter functions active.

**Figure 3.17** Most parametric equalizers will have gain, frequency and Q controls.

A wide Q will allow the equalization (EQ) to cover a wide range of frequencies, while a narrow Q will allow the user to hone in on a particular feature of the sound. The shape of the bell is also known as the *slope* of the bell curve. The slope is the rate at which the level drops above the peak frequency measured in dB/octaves.

A *Shelving EQ* is another equalization pattern or curve that is useful in audio manipulation. A shelving EQ is where the boost (or cut) extends from the chosen EQ frequency all the way to the extreme end of the range.

*Gain* or cut is the degree of boost or reduction at the center frequency measured in decibels (dB). Reduction at the center frequency produces a visual curve that dips downwards, forming an inverted bell shape. EQ cut is vastly underused by many sound mixers. The gain and Q will determine the slope of the bell.

Figure 3.18 shows a screen display of equalizer processing.

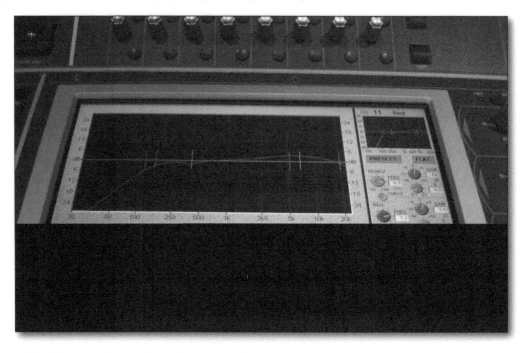

**Figure 3.18** Screen display of equalizer processing.

In a dedicated function-control mixing console, like the SSL 4000 and most analog mixing desks, each channel has a two- or three-band parametric equalizer configuration with variable high mid-frequency (HMF) and low mid-frequency (LMF) sections that has controls for frequency, gain and Q (Figure 3.19). In a digital console, it is common to have a four-band parametric equalizer; however, some digital consoles have DSP (digital signal processing) that is completely programmable.

**Figure 3.19** SSL screen display with equalizer curves, dynamics and VU levels on auxiliary sends.

## Filters

A filter is the simplest form of EQ. A filter is a fixed equalizer setting usually for a specific audio effect. A narrow bell-shaped curve is also know as a *notch filter*. A notch filter has a very steep bell curve, causing the process to affect very specific frequencies. A notch filter evolved from the need to remove electrical hum or air-conditioner noise.

### Low-pass Filter/High-pass Filter

The low-pass filter allows low frequencies to pass but reduces the level of high frequencies. A high-pass filter reduces the level of low frequencies generally below 80–100 Hz.

### Bandpass Filter

A bandpass filter is a notch filter that affects a very narrow band of frequencies. For example, it is common for microphones to receive excessive amounts of low-frequency energy from wind rumble and these filters are useful in cleaning up the mix.

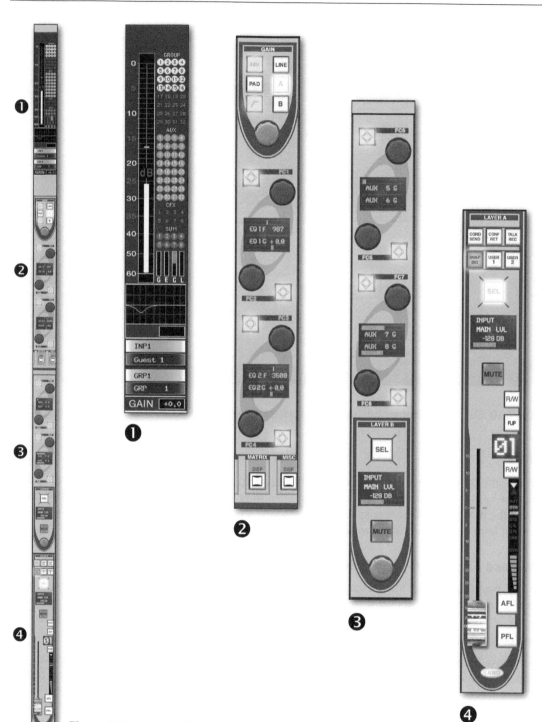

**Figure 3.20** Lawo multifunction control channel strip.

**Figure 3.21** Close-up of function selection control in Euphonix mixing console.

**Figure 3.22** Close-up of function control knobs.

## Analog vs. Digital

In the channel strip of the Solid State Logic 6000 and other analog mixing desks, the equalizer section can be assigned to either the mix channel or the monitor channel, or the dynamics section for side-chain applications. This is one limitation to an analog mixing console where the equalizer can only be used by one signal path. In a digital design, the equalizers, dynamics controls, and delays can be inserted on any channel at virtually any place.

The dynamic selection puts the equalizer in the dynamics section, which is useful for "de-essing" or sibilance reduction. Many digital mixing desks offer a de-essing preset in the equalizer control section. A parametric equalizer can be adjusted to perform the same function as a "notch filter" or shelving EQ when the Q frequency bandwidth is properly set to wide or narrow and the gain increase "cut" or reduction is appropriately applied.

Shelving EQs, filters and on/off switches are perfect "soft switch" functions for digital consoles because of the flexibility of programmable functions. In an assignable function control design, one set of control knobs can control the equalization for a bank of channels or even all channels in a mixing desk.

### Equalizer Tip

*Use the EQ to boost and cut specific frequencies. If a mix is congested, it is very likely that too many sounds or instruments are competing for the same frequency bands. Solo each track and use the EQ to ease off frequencies that are not really important for that specific sound source. Some tones are important to the reproduction of certain frequencies and some frequencies should be boosted.*

*Use the midrange EQ boost set to a very narrow Q to sweep the frequencies to identify the frequencies where there may be a problem. The reason why the frequencies are boosted first is that it is easier to hear the problem frequencies than if you cut.*

**Figure 3.23** Dynamics section. Yamaha mixing desk. The basic threshold, ratio, and release controls are similar on all dynamics control devices. Some designers add an attack control to the circuit. Threshold is the level where the compression is activated.

# Dynamics: Compression Controls

Rotating the control knob clockwise increases the amount of compression from 1:1 (no compression) up to infinity. The "1" represents no increase in output level, regardless of input level increases above threshold.

The *compressor* evens out the difference between the loud and quiet parts of a sound envelope by crushing the audio if it gets too loud and raising the audio in the quiet sections. A compressor/limiter will have control over the *threshold, ratio, and release* or *decay*.

The threshold determines how high the signal must reach before the compressor kicks in. The ratio sets how much compression is applied as the level rises above the threshold. The release adjusts how fast the compressor lets go once the input signal drops below the threshold. Some consoles will have an adjustable "attack," which sets how fast the compressor kicks in once the threshold has been reached.

The *stereo link* switch allows the two sides of the unit to be linked together for processing stereo signals. Some processing units will have output controls, which set the output signal level.

### Assignable Control Functions

*Control functions for the equalizer and dynamics sections tend to use similar control operations. Equalization often uses four variable-frequency ranges with a level and Q control for each band. The dynamic controls also use a group of four control knobs for threshold, attack, ratio and decay. With a digital platform, a set of control knobs has random access to channel and control functions and can switch instantly to operate on any channel and function desired. Reassignment of the control knobs between the different processing operations permits multipurpose usage of the control knobs with a certain level of familiarity.*

*There are console designs that place a single set of controls for equalization, compressor, limiter, gate, expander and auxiliary sends in the center of the console (Figure 3.24). Plasma screens display the changes for instant reference of the processing, routing and metering.*

**Figure 3.24** This mixing console uses four assignable control knobs to adjust all the audio parameters in the channel module.

**Figure 3.25** Some console designs place multifunction controls on each channel strip. This desk uses eight concentric control knobs whose functions are selectable by the row of buttons below.

## Auxiliary Sends/Effect Sends

Entertainment and sports audio require a variety of mixes to satisfy production demands. Auxiliary sends, also known as effects sends, are a flexible audio signal path that can be used to combine signals for routing or processing. Broadcasting uses auxiliary mixes for recording specific audio, for talent IFBs (interruptible fold-back) or production pre-hears.

Sports uses a lot of video replays in which complete and proper audio is essential to the coverage. A golf tee may have up to four microphones that need to be blended together for appropriate audio playback. Each microphone channel will have an auxiliary/effects send pot to adjust or mix the relative strength of that microphone in the new mix or auxiliary mix.

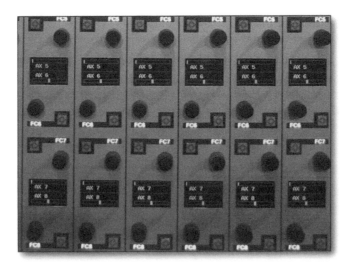

**Figure 3.26** Assignable function control set for Aux 5-Aux 8.

Individual and specific mixes are often needed by on-camera talent. The IFB is a mix that is used by on-air talent to hear themselves and other essential audio and is the means by which a show's producer can talk directly to talent. (See Chapter 4 and Chapter 5.) *Production pre-hear* is the means by which you can preview audio before it goes to broadcast. This is common with sideline reporters who will work with the sideline producer on a story or interview.

This is suitable for foldback because a mix can be constructed completely independently of the positions of the mix faders. Auxiliary sends can be either *pre-fade* or *post-fade*, which is useful when sending signals to be recorded.

Pre-fade means that the signal is taken from a point before the fader. Post-fade is suitable for reverb since, when you fade the signal out, you generally want the reverb to fade out as well and be in proportion at all other fader positions.

While each channel send pot is designed to set the relative strength of the input signal, the overall strength can be conveniently controlled by the effects send master.

# Faders

On an analog console, the mix faders control the balance or mix of the audio levels to the groups and to the mix bus. In-line audio consoles will have either one or two mix faders. In a single fader configuration, there is one long throw fader, while in a dual fader configuration, there is a single long throw fader and a shorter fader. Typically, the large faders have microphone preamplifiers and phantom power, whereas the small faders are normally a line-level input. Small faders may be called *monitor mix* or *tape returns* and normally control the level of the multitrack tape returns signal. Remember the origin of most modern mixing consoles was the recording studios and many terms are still used. In broadcasting, the monitor mix faders are very useful for playback sources such as video, music and sound effects. I have seen some mixers use the small faders as mix re-entries, as one way to organize the signal flow and keep key elements of the mix within arm's reach. (See Figure 3.27.)

In an analog mixing console, the mixing fader is after the microphone preamplifier, equalizer, dynamics and essentially post any gain control. (The preamplifier, equalizer and dynamic control can either boost signal level or reduce signal level.) The mix fader itself cannot boost the signal level, but can only reduce it. The mix fader has no influence over whether an individual signal is distorted or not. If the signal in a particular channel is distorted, usually the input gain setting is too high. Lowering the fader will simply lower the audible level of the distortion and only by proper gain structure will distortion be eliminated.

**Figure 3.27** This Calrec mixing desk has a single mix fader per channel slot with four control knobs to adjust all audio functions..

The mix faders are the closest controls to the audio mixer and are used most often during the course of a show. There are analog and digital mixing-console designs with single and dual in-line faders providing tremendous capacity in smaller packages. Digital consoles offer the tremendous advantage of universally assignable faders. This means that any single audio source or sub-mix of an audio group or master mix can be assigned to essentially any fader.

# VCA Grouping

Fader automation arrived around the mid 1970s with the introduction of the *voltage controled amplifier* or VCA. Most large consoles offer a function known as VCA subgrouping. Subgrouping is a technique by which several channels can be routed through groups, which are then in turn routed to the main outputs. The VCA allows control of many input sources by a single submaster fader. The VCA does not degrade the signal. For example, the eight plus channels of an audience mix can be balanced to give a good blend and then routed to VCA faders for convenient overall level control without disturbing the balance of the individual faders.

On a console with VCA grouping, the channel faders have a small thumb-wheel switch that can select one of usually eight VCA masters. A thumb-wheel switch is a rotary switch that you roll forward or backwards to select the VCA group that you want. When a VCA master is selected, it controls the overall level of all channels set to that group. So, the level of the entire group is controlled by one fader. With conventional subgrouping, the audio is mixed through the group fader(s). With VCA grouping, the VCA fader simply sends out a control voltage to the channel faders.

**Figure 3.28** SSL dual fader configuration.

VCA mixing is useful when you have a set of faders that have a good relative balance between each other, but the audio mixer needs to have control over the entire group of faders. This is common with an audience mix and some mixers use a foot volume control pedal with a built-in VCA to control that cluster of faders.

Another common use of the VCAs occurs when the mixing console is so dense and long that it is not practical to mix on.

Automated mixing is a function used predominantly by the recording world. Automatic recall is very useful for a show where you have to recall the mix of a musical act or if you have a recurring set-up.

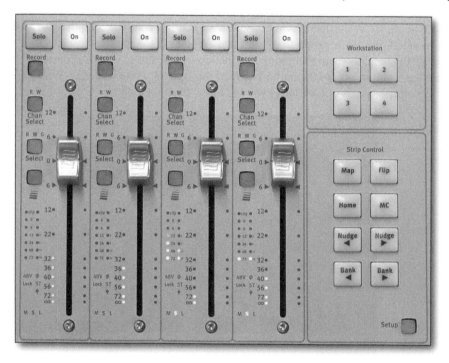

**Figure 3.29** Euphonix faders with LED (light-emitting diode) displays for pre-fader level monitoring.

Audio-follow-video has been used for years but was limited by the mixing console or the interface. The video switcher can trigger a switch called a GPI which opens or closes a circuit. This is very useful for cutting a fast succession of camera shots where the audio always has the same blend of sounds. Digital consoles are capable of making a "snapshot" of a particular combination of audio for instantaneous recall.

**Figure 3.30** Lawo assignable faders. In many mixing consoles the faders can be assigned for microphone, line, group or VCA controls.

# Routing the Signal Path

A mixing console will have many inputs and a variety of mixable or assignable outputs such as auxiliary, group, matrix and main outputs, also known as "master" or "mix" outputs. The assignment and directing of audio signal flow from an audio source to a particular audio destination is known as *routing* the audio signal.

A simple example of signal routing is the use of a direct output or buss assignment switch to direct the audio to another destination. The *panorama* or *pan pot* normally controls the left and right side of the stereo output but often controls the odd/even assigning of stereo groups or stereo auxiliary outputs. By turning the pan pot all the way to the left, the audio signal is sent directly to the left output of the mixing console. Often the routing switches and panpot work together. For example, the routing switches will direct the audio to Groups 1–2, or Groups 3–4, or Groups 5–6, or Groups 7–8 and the main stereo ouput while the pan pot will vary the level between the left and right or the odd and even audio channels.

All these outputs are mixable outputs of one or more audio sources, where a direct output is the routing of usually a single channel of audio that is not mixed.

A professional analog mixing console might have 48 inputs, 8 group outputs, 8 Aux outputs, a matrix section and 2 main outputs. In most configurations any channel or number of audio channels can be routed to any number of the groups or auxiliary sends plus main output left and right.

*Patching* is another method of signal routing that uses a hard wire to move the audio signal from a source to a destination. Virtually all broadcast installations will terminate the inputs and outputs of a mixing desk to patch points in the audio control room.

### Spatial Orientation

*The pan pot is the basic imaging control for a mixer to take advantage of spatial placement of mono audio sources. In early stereo mixing consoles, the choice of placement of a mono signal in the stereo soundstage was simple—it was either left, center or right.*

*With the invention of the panoramic potentiometer, known as the panpot, you can position a mono signal anywhere in the stereo soundstage. If the panpot is designed properly, then the sum of the levels in the left and right channels will remain constant whether it is panned left, center or right. You should be able to pan all the way from left to right, and the signal level should be subjectively the same all the way. When a signal is panned center, it needs to be 6 dB lower in level than when it is panned hard left or hard right. (See Figure 3.31.)*

**Figure 3.31** Surround panning with an airplane-like "joystick."

*When planning the stereo image, placing mono sounds across the full range of the stereo field should be considered. A stereo input source coming into two channels may be panned hard left on one channel and hard right on the other to maintain the maximum stereo effect. The problem is compounded in surround sound. Having no sounds panned hard to one speaker or the other tends to make a listener aware of the 180-degree sound field rather than specific locations.*

*Designing a 5.1-channel panpot is a far more complex affair than stereo, and similar considerations regarding levels apply. A 5.1 panpot might have to pan from left to left surround, center to right surround, perhaps even into the low-frequency effects channel. All the channels need to contribute to the acoustic signal in the correct proportion.*

# Channel Insert points

The insert point of the mix channel is used to route a direct audio signal from that specific channel to another destination. The channel insert point comes directly after the preamp stage of the channel and in some consoles it comes after the equalizer, which can be better for compression. The insert send is often used to send the signal to an external processor such as a compressor, noise gate or equalizer. When a processor is inserted into a specific channel, the process will be specific to that channel only. The insert return accepts the output of the external unit back into the channel and completes the signal path. Similarly, a reverb unit connected to the insert point will only affect the signal on that channel. If the reverb is inserted into a group, then the process will affect everything assigned to that group.

There is another use for the insert point, besides sending the signal through an external unit. The insert send can be used by itself as an extra output from the console for that signal. However, you would normally only do this if you were already using every other output. To use the insert point in "send only" or "half-normal" mode, the patchbay is wired so that it links the tip and ring of the jack in order to pass the signal through, as well as providing the additional output. Another way to accomplish this is to use the insert send and go to a distribution amplifier (DA) and patch the output of the DA to the channel insert return, thus completing the signal path and giving you multiple feeds of that signal from the DA.

When using the insert sends, a switch in the channel strip may be needed to activate the insert point while the audio passes straight through to the rest of the circuitry.

# Mixing Console Outputs

The number and flexibility of the outputs of a mixing desk is a true measure of the functionality of that console. A basic mix desk will have a stereo output, auxiliary outputs and several group outputs, and if it is a surround sound mixing desk, of course it will have a surround-sound output as well.

Each input channel can be or not be assigned to either or all of the outputs. For example, the audio signal can be assigned to the stereo output and/or surround mix outputs, and/or the auxiliary outputs and/or the group outputs. Even though the individual audio channels can be assigned directly to a mix, it is common for a mixer to assign similar audio sources to groups and balance groups of audio to a mix. For example, you would assign all the announcers to a group or all the drums to a stereo group and then assign the groups to the main mix. If the overall level of the announcers or drums needs to be adjusted, then you only need to tweak the group level and not all the individual faders.

**Figure 3.32** Digital break-out box, commonly called a *stagebox* or *fieldbox*.

When an audio channel is assigned to a group, it is then combined, summed or blended with other channels within that group. Groups are usually arranged in stereo pairs, often with a pan pot to adjust the level between the odd and even groups. Additionally, groups will have a direct output and usually a channel insert. Analog mixing consoles usually do not have signal processing built in and use channel inserts to send the channel signal to a processor and return the processed signal back to the group.

The problem with routing channels to groups is that only a single mix of the groups is available when the groups feed to the stereo mix bus. Yamaha introduced an additional summing matrix that allows the groups to be mixed together to another output. A typical matrix usually has 10 inputs, one for each group plus left and right.

The volume level of each of the ten inputs is combined and summed to a single output level control, which is essentially an 8-by-1 mixer. Since the matrix is at the end of the signal flow after the groups, a designer can add as many matrix channels as desired. The matrix section is useful for creating multiple foldback mixes for IFBs and control-room monitoring that may require a different mix.

# Monitor Selection

Audio monitoring is a critical function for the sound mixer and is often very difficult in the confines of the television truck. Most mixing consoles provide internal switching between monitoring speakers and monitoring sources. The monitor speakers will have a level control that sets the volume intensity of the main control-room loudspeakers. The output to the control-room loudspeakers is normally derived from the console's stereo output or surround output, but may be switched to external input sources.

The external or ancillary input can take their signal from any one of a number of internal sources or externally patchable sources such as distribution amplifiers. Many sound mixers monitor the transmission signal to insure the audio going to the satellite or fiber optics is correct and properly calibrated.

There are many mixes going on in the average television production, making signal monitoring a critical function to insure the proper audio feed is going to the proper destination (Figure 3.33). The broadcast will have a stereo and probably a surround mix of all the sound elements including voice, atmosphere, music, sound effects and sound for video replays. In a sports production, it is common to feed the announcers a mix of all production elements minus the sports sound effects. The announcer may be at the sidelines or in the pits and the additional sports sound is not necessary and may hamper the intelligibility of the audio. It is absolutely necessary for the sound mixer to be able to monitor the various mixes to judge their integrity and proper content.

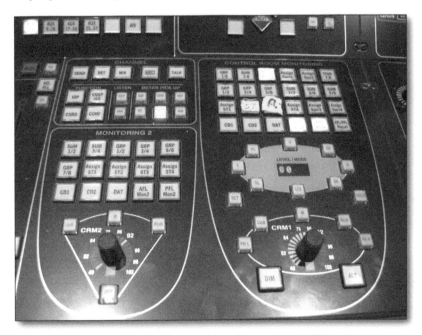

**Figure 3.33** Monitor control section. Source selection and level control. This selection panel routes the desired signal to the speaker monitors in the audio control. An adjacent and identical set of controls routes the audio signal to production control room.

Monitoring the various mixes and feeds is essential, but being able to monitor the actual signal being broadcast is critical. There was a common practice for the transmission of the television signals to be controlled by an audio/video router, to permit the television truck to finish production and rehearse without the pictures and sound being transmitted. The engineer in charge would leave color bars and tone on the transmission path and then make the switch to the television production at the proper time. Most routers are programmed for the audio to follow the video switch, but this time the audio stayed latched to the tone! The poor sound engineer had to patch around the problem and maintain a sound mix because the program was being recorded with the audio directly from the mixing desk.

Many monitor control sections will be able to perform useful functions such as to dim and cut the monitor speakers plus combine the stereo signal to make a mono signal. The *dim* button reduces the monitor level to a preset volume, while the *cut* button turns off the control room monitors. *Mono* sums left and right channels to the monitor, allowing the degree of mono compatibility of the mix to be assessed.

Many mixing consoles allow the operator to switch between different sets of speaker groups: stereo, surround and other types, such as nearfield speakers.

**Figure 3.34** This audio control room has an additional pair of near field monitoring.

The monitor section is generally used to scrutinize everything from the show mix to the "in-ear" monitor mix. It is often necessary to select and monitor a particular element of a sound mix without changing the audio's relative contribution to the mix. Mixing consoles have the ability to *solo* any channel so that you hear the individual channel by itself through the monitors. In broadcasting, this solo must not affect the signal being amplified or broadcast.

The *prefade listen* means that you hear the signal of the selected channel alone from a point before the fader, so that the fader level has no influence on the PFL level. On some consoles, the PFL signal is routed equally to left and right speakers, while on others the position of the pan control is retained. PFL is often used for setting the gain control on each channel. One at a time, the PFL buttons are pushed in order to set the gain so that the main meter reads a good strong level without going into the red.

### Speaker Monitoring

*The television truck is a difficult environment in which to monitor the audio mix. There are not only distractions from the audible cues of the director and producer, but the OB van is also a noisy environment from air conditioners, equipment cooling fans, and thin walls. Speaker placement is very difficult because every piece of equipment in an OB van competes for space in a situation where the OB has size and weight restrictions in order to be mobile.*

*Speaker placement and monitoring is even more critical for surround-sound, particularly when monitoring matrix-encoded mixes. The center channel is the dialog channel in a surround-sound mix and loudness levels of dialog continue to be a problem. One issue that complicates the problem of getting a balanced mix is proper speaker placement and speaker distance from the mixer.*

*As mixing consoles have gotten wider, adjusting a mix has become more difficult for the operator. Many digital consoles provide a set of function controls in the middle of the desk directly between the loudspeakers, so the engineer can remain in the "sweet spot" no matter what channel is being adjusted.*

## Digital Mixing Consoles

The basic function of a mixing console is to combine and blend audio signals—essentially audio signal management! The problem of capacity increases in surround because of the increased number of single channels required per sound source. Essentially six single channels are required for each surround output and sound playback from video machine, audio effects and music sources were previously stereo or mono and are now 5.1. A large high-definition production may have up to 30 full surround channels, plus as many mono and stereo.

With this level of complexity, digital consoles are essential because of their capability of managing large numbers of inputs and outputs in a relatively small package. In a television truck, size and weight of the mixing desk is an issue and the size and capacity of many analog mixing desks have reached diminishing returns.

**Figure 3.35** Early designs incorporated digital controls with analog inputs and outputs and all the internal processing and routing of the audio was analog. To get the functionality in the signal flow, a new design needed to have random control and assignability of the signal flow and digital audio was necessary.

The answer to these problems has been the *digital assignable console*. In an assignable console, usually there are channel strips with a large mixing fader just as before, but redundant function controls such as equalization and dynamics are shared between channels (Figure 3.35). Functions such as EQ and dynamics do not need to be replicated for every channel where one set of controls can be placed in the center of the console or a set of function controls can be positioned with groupings of channels faders, as seen in Figure 3.36. Instead of fumbling between skinny channel strips searching for the right control, now only the very essential functions are available for instant access.

**Figure 3.36** Channel strip with selection controls for multifunction controls.

One of the best features that many mixing consoles offer is screen visualization and touch sensitivity. Screens offer instant visual references of processing, such as EQ curves and dynamics as well as mapping of the signal flow and sound design. Touch-sensitive control knobs follow the operator's work flow and speed up the processes of adjusting a mix. Touch screens display several settings and a single touch on any element in a channel strip, opens expanded view windows for advanced functions.

With an assignable console, the mixer has the freedom to assign any source to any position on the console, saving it as a layout. This includes knobs on the channel strips and pots used to control EQ, dynamics, compressors, gates and auxiliary sends. Digital consoles have no physical link between a control and its associated electronics, which allows more flexibility for control layout.

**Figure 3.37** Single fader design with display screen and control functions.

Designing an ergonomically effective surface has been the challenge of mixing-console designers. In an assignable console, there are fader strips, screen displays and a central control area that operates global functions such as the monitoring of the audio desk.

There are two basic functional designs for most assignable mixing consoles. A common approach is for the digital console to look and feel like a traditional in-line analog console with a vertical strip of control knobs in-line with each fader. Each channel strip includes a set of multifunction control knobs that control equalization, dynamics and routing.

A second approach is to have a group of fader control strips that use a common set of central control functions for equalization, dynamics, and routing. This allows more faders in a compact space and a central processing unit (CPU) for a physical group of faders. The central processor has a physical control for each parameter that you can adjust in the strip path, sometimes duplicating controls. Operation of the console is spread between the central processor and the channel strips. (See Figure 3.38.)

**Figure 3.38** The modern digital console has controls that look pretty much like conventional control knobs except the operational functions change as selected. For example, when the EQ function is selected, the controls function like a band parametric with individual controls for frequency, gain and Q for each section. Additionally, an EQ section may have high- and low-pass filters and a notch filter.

Most digital consoles have a function indicator on each strip, which shows the functions you have assigned. Typical channel functions on a digital console include: input gain and trim, input select, pan or width, equalizer, filter, dynamics/compressor, limiter, expander gate, aux select, aux sends, signal routing and movable insert points.

Most mixing consoles have a central area where *global control functions* are managed: loudspeaker levels, master solo controls, main console output controls, main aux/cue controls and levels, keyboard and talkback, oscillator. Monitor select sections will usually have patchable inputs for the mixer to create its own monitoring chain.

In many digital consoles, channel strips can be configured to operate in mono or stereo, automatically linking left and right paths in stereo and providing a meter display of the left and right signals.

Many digital console controls do not have any endstops so the knob goes round and round, if you want it to. The position of the knob is shown by a ring of LEDs, which can illuminate individually to show position or indicate other functions.

## DSP Processors

*DSPs or digital signal processors provide the basis for all digital mixing consoles. The DSP chip is a microchip or set of microchips that processes, routes and performs all the mixing functions of the digital audio. DSP chips as with all microchips, advance significantly every couple of years and 2006 chip technology offers 480 equivalent mono signal paths with signal processing capable of full EQ and dynamics to all channels. This single chipset processor is capable of being configured to surround sound for 78 x 5.1 surround channels with full signal processing. Output configurations are programmable for 24 stereo groups or 8 x 5.1 groups with full EQ and Dynamics, 4 x main outputs, 48 multitrack outputs and 20 auxiliary sends. System redundancy of all processing elements is provided through a second DSP card.*

*In a small or medium size mixing desk, a single DSP chip can control the general signal processing and routing, but there can be problems with high channel count and high sampling rates, such as 96 kHz and 192 kHz. DSP processors have a fixed capacity and as you increase your sampling rate, you increase the load on the DSP circuitry. With the DSP set for a sampling rate of 44.1 kHz, you may be able to open all the EQ and dynamic functions on all 72 channels of the mixing console, but at 192 kHz, you may only be able to operate 36 channels. Control and sync information is impeded, resulting in diminished performance.*

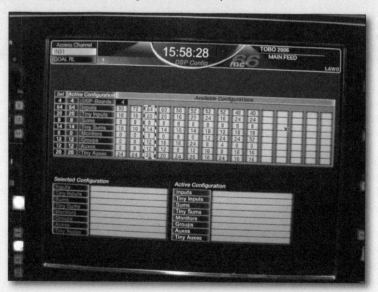

**Figure 3.39** DSP configuration and status screen.

*Digital console design incorporates DSP processing that is scalable so that additional processing capacity can be accomplished by adding additional DSP processing chips.*

*Additional DSP provides additional capacity or redundancy.*

Audio and video sync have been an issue because audio and video are sometimes captured and processed along separate signal paths. This can cause synchronization or "lip-sync" problems, first noticed in the 1980s when wireless microphones became prevalent. In HD productions it is still necessary to use standard definition (SD) equipment and the "up-conversion" of video signals introduces delays that must be compensated for. Digital mixing consoles and digital processing equipment can also introduce delays from analog-to-digital conversions.

Digital consoles allow the user to change to signal flow blocks, which is extremely useful when it is necessary to use the delay section of a console to resync the audio to a camera that has been processed.

Digital consoles offer the greatest future potential and flexibility in sound design and mixing. As previously noted, VCAs are extremely useful when a single fader is needed to control a relative mix of multiple faders. The digital console will permit combining and mixing unlimited faders and audio sources into individual mixes that are independent and unaffected by any other mix.

**Figure 3.40** Central processing section with layer selection switches.

The digital consoles provide the ability for the sound mixer/designer to blend a dozen different sound sources and output a mix with surround perspective and be able to assign that mix to a single fader for mixing. For example, an over-the-shoulder close-up of a drummer in concert setting may call for the producer to want to push parts of the drum mix. A fader can be assigned a specific blend of microphones for one shot and a second fader can be assigned a different mix of microphones and the sound mixer can even vary the blend of just those two faders for a transition. The key concept is that, in the new mixing world, a single fader does not represent a single sound.

The number of channel strips can be reduced as well. For instance, 48 fader/channel strips can be assigned to cover as many channels as the console has electronics for. Possibly as many as 96 or 144 separate channels are created through layers or banks. Layers exponentially increase the mix capacity and are useful for grouping sections of the mix that are "stems" or sub-mixed components of the sound mix. For example, Layer 1 is the layer where you balance group mixes and overall production mix; Layer 2 is the music or playback sources; Layer 3 is the surround atmosphere microphones, and Layer 4 is the event-specific microphones.

There is a downside to digital mixing consoles. An analog console is generally fairly easy to operate and there is a learning curve with digital consoles! Also, every digital console is different and has its *own* learning curve. Operators need to learn new functions and operations quickly and this has driven designers to study existing architecture, signal and work flow and design familiarity into the new work surfaces.

Additionally, with any assignable console there is likely to be a problem that some of the settings you have made are currently hidden in layers. Also, assignable controls can be accidentally changed, since you are now using the controls for another purpose. Display screens make it possible to view the most important settings in an abbreviated form.

Audio synchronization through the system is a significant concern. Sometimes audible delay occurs when moving the audio signal through different signal paths in a digital board. Every analog-to-digital conversion of the audio and video will cause delays. For example, a digital console may have analog insert points that will cause a delay if you insert your favorite analog compressor.

## Outboard and Processing Equipment

As versatile as the mixing desk is, the audio mixer will be required to operate additional equipment. The most common processing device in broadcasting is the compressor/limiter. While most digital consoles have dynamic functions built into each channel and usually on group and main outputs, there are still many analog consoles in circulation that require "outboard" compressors like the DBX 160 series or Aphex Compellers.

The DBX 160A and the Aphex Compeller compressors have set the standard in broadcasting for many years. The DBX is easy to use with only three controls—threshold, ratio of compression and an output level control to make up any reduction in level. The DBX 160A features gain reduction monitoring and switchable input or output metering.

**Figure 3.41** Calrec mixing desk with DBX 160A compressors, Digi Cart and phase scopes in over bridge.

It is common for the final output mixes to be processed by an additional dynamic controller. The Aphex is popular with broadcasters because of its inaudible compression and increased system gain without constant "gain riding," which is common with sports announcers. The Aphex uses different terminology for similar functions. The *drive level* sets the desired amount of processing, the *process balance* controls between leveling and compression, and there is an output level for unity gain. The Aphex uses different detector circuits for leveling, compression and peak limiting.

Electronic processing of the surround audio signal cannot be performed adequately by using mono or stereo tools and requires a dedicated multichannel processor. When mono or stereo compressors are not linked, dynamic sound variations in the individual sound channels can cause the sound image to shift. Television generally treats the left and right front channels as a pair, while the surround channels are treated as a pair with stronger compression in order to hear the surround channels at lower listening levels. The center dialog and low frequency effects (LFE) channels are generally processed separately because of the tremendous dynamics.

This is a six-channel device with proper channel interaction and a close connection between the channels that is needed particularly when upmixing from two channel sources to five channel mixes and downmixing from surround to stereo.

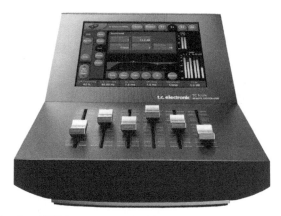

**Figure 3.42** The TC Electronics 6000 is a programmable multichannel processor with dynamics, delays and even a surround sound "unwrap" program.

All music and sound effects are currently played off a Digi Cart. This is a hard-disk recorder and playback unit that offers remote triggering and limited editing function. Assignment can be stored to hard disk and recalled whenever necessary. Magneto-optical and floppy disk drives are provided for removable storage. The most useful feature of hard-disk recording/playback is instant access to the music or effect. The predecessor to the Digi Cart was a tape-based endless loop cart and cueing was a nightmare. Many sound mixers like to have the remote control pad close, because the Digi Cart is used for bumper music. Most sports productions require a minimum of two Digi Carts, one for triggered effects and the second for music. ESPN uses two different Digi Carts with separate music tracks for different shows. (See Figure 3.43.)

**Figure 3.43** Digi Carts are used for sound effects and music.

Some video recorders can be played back as individual channels of audio or as a Dolby E stream.

**Figure 3.44** Aphex remote control microphone preamplifiers.

Entertainment and music programming is usually done in front of a live audience and requires a higher level of sonic clarity than most sports programs. Microphones are shared between PA, monitors and broadcast. Microphone signals may be split several times, which degrades the signal. Additionally, long mic level capable runs are susceptible to line noise and electromagnetic interference, which is boosted at the end of the lines by the console microphone preamp. By placing microphone preamplifiers close to the mics, the need for long mic level runs are eliminated. The Aphex provides five simultaneous outputs (two analog and three digital), which eliminates the need for splitter boxes.

All live applications require the monitoring of mic preamp gain and the Aphex provides remote adjustment and function monitoring through a PC running controller software. Using PC controller software, the Aphex has a screen that displays all parameters and metering of one unit at a time, with a capacity of up to 16 units. All channel status information and metering are displayed in real time and a channel can be selected and settings modified. Scenes can be saved, modified and recalled within the control software. The software also contains a "learn" function in which the channel(s) can adjust the preamp gain to a definable peak value based on the input level during the learn time.

## Organizing the Mixing Desk

The mixing desk needs to be organized so that critical sources are within an arms length and the speaker monitoring is accurate. The inputs to the mixing desk should be organized in stereo groups of related audio such as announcers, event sound, and music and video playback. Each group should be intuitively placed along the foot print of the mix desk so that it is comfortable for the mixer to find each audio source. For example, a left-handed A1 needs to have sources that can be constantly adjusted within the reach of the left hand.

The first layer or grouping of inputs is announcers and all voice sources. The next groups of inputs to the mixing console are usually music, sound effects and video playback. These sources require two-channel stereo inputs to the mixing console. Triggered sound-effect playback seems here to stay, and once the playback device is loaded, programmed and tied to the video switcher, which triggers the playback, there is little to worry about.

The final level of inputs to the mixing desk is the sound effects group. Sound effects come from the camera microphones, spot microphones near the action of the event and atmosphere microphones placed around the facility.

Clear labeling and documentation of signal flow and patching is necessary in order to troubleshoot in the middle of a broadcast.

The layout of the mixing desk is a personal preference and will change with different events. The most radically different layout I have seen is Denis Ryan's effects mix for NASCAR. Most of the odd-numbered cameras are to the left and even-numbered to the right, with the in-car microphones in the middle between each group. The logic is the back and forth from left hand to right hand of the camera cuts; any hand can get to a quick in-car cut.

**Figure 3.45** This mixing console has very clear labeling on the mixing faders as well as the meter bridge for quick identification.

# Cleaning a Mixing Console

Mixing consoles attract dust and dirt and, with all those knobs and buttons, it is very difficult to clean. Three kinds of dirt are prone to collect in the mixing console: dust, smoke and finger marks. The problem with dust and smoke is that they enter the moving components—the faders, potentiometers and switches. These are devices that rely on good electrical contact through surfaces that barely touch. Separate them by a dust particle and you get the familiar scratchy sound of old equipment. The better varieties of these components are sealed more effectively, but they cost more. Cigarette smoke is known to wreck a console after a couple of years.

Grease from sticky fingers isn't a problem in itself, but it makes the dust cling. Consoles benefit the most from regular cleaning, so dust doesn't have time to form a stuck-down layer. Stickiness from spilled drinks is even worse, although that can be avoided by keeping drink trays below the level of the console, so they only spill on the floor.

To clean a console, you need a tool that will get in the crevices. A vacuum cleaner is ideal if a very small nozzle is used. Vacuum it daily and your console will be spotless. Faders and potentiometers will be smooth and not scratchy. If dust collects and sticks, you will need something more aggressive, such as a 25-mm wide paintbrush. When dusting the console with a brush, try not to brush the dust into the components but away from them, particularly the faders.

# 4 The Audio Assistant

The role of the audio assistant is to work with the senior audio person and television crew to set up, maintain and operate all the audio equipment at the television event. As a general rule the audio assistant takes care of equipment outside of the television truck while the senior audio takes care of the set-up inside the television truck. The duties of the audio assistant vary with the type of production and most audio assists have a wide range of skills and experience (Figure 4.1).

The process begins with the ESU (engineering set-up) when crew members install either copper or fiber cable between the television truck and all broadcast positions then assemble and connect the equipment. For a sports event, the audio assistants will set up the announce locations, microphones on the field and cameras, plus all the communications between production, engineering, scoring and event officials. At an entertainment event, the audio assistants will rig microphones, prepare radio microphones and install a very complex communication network.

The sports audio assistant will set up and maintain multiple announce locations in the press box, at ringside or on the playing field, wherever the announcers will voice the action of the event or conduct interviews. Often, there will be a separate dedicated room for the television and radio announcers to work in. This announce-booth configuration is very common at stadiums for football and baseball and is usually in the vicinity of the other media or press rooms. Contrast this to basketball, where the announcers are at a court-side table. There will always be an audio assistant wherever there is an announcer, including the "roving talent" at golf and motorsports, where the audio and maintenance functions are performed by the radio technician.

The audio assistant will place all microphones on the field of play, on the cameras and at specific locations designated by the audio mixer. Most microphone placement has been thoroughly tested and listened to, and often the audio assistant will find that a sound mixer can be very specific about microphone placement.

**Figure 4.1** Audio assistants place microphone at corner of gymnastics mat and tape down the cable for safety and a clean look.

A significant responsibility of the audio assistant is to install and maintain the communications network. There are communications requirements between the producers, director, announcers, and production support crew as well as engineering communications between the operators and technicians. The telecast cannot function without clear, intelligible communications and installation and maintenance is a critical part of the job duties performed by the audio assistant. At larger events, multiple audio assistants will be in the field, on the set, at the announce booth and in the broadcast compound. The senior audio in charge or A1 will assign the audio assistants (A2s) to perform different parts of the set-up according to their skills and personalities.

## *Audio Set-up*

The preparation begins with an engineering meeting with the senior audio person who will provide assignments to the audio crew along with a *mult-sheet*. (See Figure 4.2 and Table 4.1.) The mult-sheet provides a good indication of the locations and function of the outside equipment. Be aware that each location may require multiple pieces of equipment to accomplish a particular function. For example, a private line (PL) position will need a communication remote operation box and a headset. The headset may need to be a single ear piece or have ear "muffs" that cover both ears. The audio assistant must understand the audio function and what gear is required along with the interface, interconnect and the signal flow.

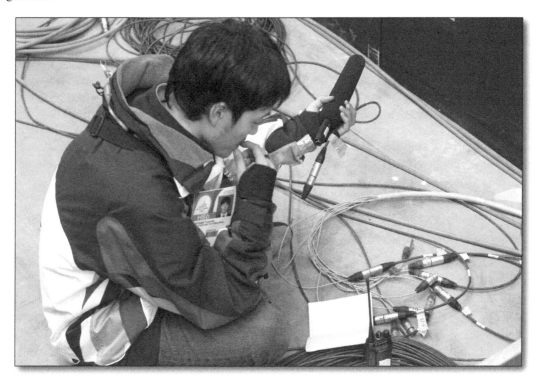

**Figure 4.2** Audio assistant checking mult-sheet to properly plug in microphone to the correct input.

The mult-sheet is the list of all the equipment and shows which pair of wires each piece of equipment will be plugged into. The *audio mult* is generally a bundle of 12 shielded wires that use a multipin connector to extend the audio mult to a greater length or to connect to a breakout where each individual wire has a connector to use.

**Booth Mult # 1**

| | |
|---|---|
| 1) Play by Play | - Headset |
| 2) Color | - Headset |
| 3) Guest | - Headset |
| 4) Hand Microphone # 1 | |
| 5) Hand Microphone # 2 | |
| 6) Spare | |
| 7) Spare | |
| 8) Spare | |
| 9) Crowd Microphone | |
| 10) Crowd Microphone | |
| 11) Crowd Microphone | |
| 12) Crowd Microphone | |

**Booth Mult # 2**

| | |
|---|---|
| 1) Play by Play | - IFB |
| 2) Color | - IFB |
| 3) Guest | - IFB |
| 4) Hand Microphone # 1 IFB | |
| 5) Hand Microphone # 2 IFB | |
| 6) Stage Manager PL | |
| 7) Statistician PL | |
| 8) Spotter PL | |
| 9) Audio Assistant PL | |
| 10) Spare | |
| 11) Venue PA | |
| 12) Talkback to Producer | |

**Table 4.1** Mult list for two 12-pair bundles of wires.

In Table 4.1, booth mult #1 is all microphones and the female connectors will be in the field with the audio assistant. Booth mult #2 is IFBs and PLs and the male connectors will be in the field with the audio assistant. The male end of the XLR connector usually denotes signal flow. For example, the microphone signal flow is to the mixing console. The output connector for a microphone is a male XLR and the input connector for a mixing console is a female XLR. The signal flow of the communications devices is from the OB van to the remote location and the output connector from the OB van to the communications box is a male XLR.

> ### Television Communications Jargon
>
> *The IFB is a single-direction listening device that the talent uses to hear a program mix, themselves and the producer who can interrupt one ear and talk directly to them.*
>
> *PL is an acronym for private line, which is the communications channel between production and engineering crew members.*

Before you take any equipment off the television truck, it is a good idea to do an equipment inventory. Since the mobile television truck travels from show to show, there is a good possibility that some equipment is missing and it is part of the audio assistant's job to bring any shortages to the attention of the truck personnel. After the inventory, collect the equipment needed for your set-up. Television productions require many parts and pieces and there is a propensity to misplace or lose equipment or even have it stolen. When moving the equipment from the television truck to the various locations, pack the parts in plastic tubs so the small parts are not dropped on the ground and lost.

The cabling and interconnect for an event will depend on the length of the cable run, complexity of the set-up and mixing-console interface. The A1 and A2 may install a couple of copper cable mults for a single-day event that has a short run, such as a basketball arena or college football stadium. Copper audio-mult would interface with the television truck through the "access bay" on XLR connectors.

Most digital consoles have either fiber or coax digital inputs and the television truck will have a field interface box with XLR inputs and outputs. This simplifies the interface and connection to a single cable. (See Chapter 3 on mixing consoles.)

Cabling is increasingly done with fiber optics because fiber is not prone to electrical interference and can carry a lot of capacity in a small, light bundle of cable. Fiber optics does not carry voltage and requires a different configuration for the communications and IFBs. Some venues are precabled and the television truck will then tap into a *house break-out panel* that is internally wired and connected from a central location to the announce booths, field, dressing rooms and various other locations. Precabled venues certainly speed up the set-up and tear-down; however, some house systems suffer from a lack of maintenance.

**Figure 4.3** This booth box speeds up the process of setting up and troubleshooting tremendously and generally makes for a neat and orderly space.

On large weekly events like football and NASCAR, where the television truck and the crew is consistent, a package like the announce booth box becomes very practical and desirable. The announce booth remote box has interfaces and power supplies for IFB and PL communications, plus amplifiers for microphone feeds to the mixing console, and can usually operate fully on a single piece of fiber cable. The booth package pictured in Figure 4.3 is connected to the OB van with fiber cable and requires a thorough understanding of the interface and set-up. There is a learning curve for the set-up and signal flow of any system! In the early '80s I set up an announce booth for ABC with an ABC homemade booth box without any documentation and consequently damaged the system with voltage connected the wrong way. The engineer on the truck was not very friendly after the incident and wrote a technical report that made me look like an idiot.

**Figure 4.4** The equipment is enclosed in a shock-resistant case and interconnected with short cables and wire looms without a patchbay.

As complex looking as this set-up appears, the goal of any package and configuration is reliability. By placing all the electronics in the announce booth and sending signals on fiber optics, the quality and reliability of the announce set-up is increased. The booth box facilitates troubleshooting because all headsets, IFBs and PLs are powered and fully functional within the booth box system. The audio assistant will be able to confirm that all systems are functioning properly, which streamlines the facilities check with the A1.

Every live television production has several positions where announcers will present the event. A sports announcer can use a headset microphone or a hand microphone and will require an IFB, or interruptible foldback. Television announcers require production support from stage managers, spotters and statisticians, which all require communications. The mult-sheet not only guides the audio assistant to the location of audio equipment but also the required purpose of the location, which directs the assistant with equipment selection to perform the function. The mult-sheet in Table 4.1 indicates that the announce booth will need three announcer headsets, two hand microphones, four crowd microphones, five IFB boxes, four communications headsets and boxes and three talkback systems. Additionally the audio assistant will need wire breakouts, cables and a small mixer to combine and balance the talkback system.

A simple and avoidable mistake is not to have the keys to the various gates or doors for each location. The technical manager will generally have or find keys as well as issue proper credentials and manage the on-site transportation, such as golf carts or other motorized vehicles. Finally, remember to take a hand radio and head for the announce booth.

The announcers wear a headset with a boom arm that holds a microphone near the mouth and lips. The microphone signal flows from the headset microphone through a switch control that has two separate paths for the audio to travel to the television truck. The normal audio path flows directly to the mixing consoles and is integrated into the sound mix. The switched path takes the announcer voice off-air so they can talk directly to the producer and television OB. It is common to amplify the low microphone-level signal to a line level before the switch, to avoid any clicking or popping over the audio path. The line-level signal is less susceptible to interference and interfaces directly with a monitor speaker or listening device for the producer to listen to. Remember, a microphone-level signal must be amplified to a line-level signal to be input into a listening speaker or most audio devices.

An announcer headset covers both ears and has a control box that is electronically attached to the television truck or studio. The system that allows the producer to talk to the announcers is called the IFB or interruptible foldback. The IFB is a two-channel discrete listening system that allows the audio technician to put a sound mix in each ear of the announcer headset and permits the producer to talk to the announcer in only one ear of the headset. The sound mix in the announcer's headset is usually a mono program feed of themselves, program effects and music, and interrupting only one ear is far less disruptive to the announcer than when the announcer completely loses their voice in the headset mix. Each announce position will normally have a separate IFB system with separate producer talk.

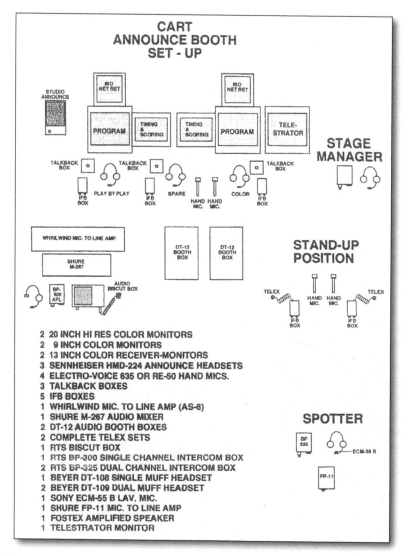

**Figure 4.5** To insure a consistent set-up with different audio assistants, the technical manager includes a block diagram of the announce booth in the engineering and production book.

The IFB is a single-direction listening device and is integrated into the matrix-based communication systems like Riedel and Telex with a separate interface and powered system. This is usually accomplished with an audio circuit and a power supply that electronically injects the IFB audio into an electrical voltage carrier. The IFB user box is an active electronic device that operates off the voltage coming from the power supply. The powered IFB and PL user boxes operate very similarly to a powered telephone circuit. Normally an IFB control box is hard wired directly to the television truck, although the IFB power supply can be located in the announce booth as in the NASCAR set-up. (See Chapter 5 for an indepth examination of the IFB control box.)

The IFB control box can be a stand-alone box (see Figure 4.6.) that is hooked to the belt of a side-line reporter or it can be integrated into a box with other functions for the announce booth. (See Figure 4.7.) The important concept is that an IFB system is the only way that an announcer hears the program, themselves and the producer. The announcer IFB control box has separate volume controls for each ear, along with different mono and stereo headset jacks to connect to. A common mistake made by an audio assistant is not using the stereo headphone jack for a headset. The announcer headset will work when plugged into the mono headphone jack, but there may be a full interrupt in both ears or only one ear will function properly. Problems and errors in the set-up will become evident during the fax check of the announce booth with the A1.

**Figure 4.6** The box to the left is a belt-pack IFB and earpiece. Notice the mono "tip-sleeve" ¼-inch connector for the earpiece. On the bottom of the IFB box, there is a mono and a stereo ¼-inch jack for an earpiece such as this or a stereo headset.

**Figure 4.7** Integrated announcer box. The button on top of the announce booth combined box takes the announcer off-air.

It has become common to use an integrated announcer control box that includes a microphone-to-line amplifier, switch control and IFB listening circuit. (See Figure 4.8.) The single box accomplishes the functions of amplifying the microphone, switching the signal into two separate circuits' paths and controlling the volume levels of the IFB to each ear piece. The single box is convenient for set-up and is very reliable because of the reduced interconnection point between the electrical functions. This box has top volume controls for each ear and the top button is for talking back to the producer. There is a 5-pin connector on this unit for a headset and a 3-pin connector seen on the front for a hand microphone as well. A single unit is easier for the audio assistant to set up and for the announcers to use. Additionally, an integrated box requires less table space. The top ¼-inch headphone jack is for mono and the bottom is for stereo.

Inside the television truck, the interrupt function is programmed into the proper user stations. Most television operations use an intercom system with programmable port that can be assigned to be an IFB audio channel. The port can be programmed to be an audio channel with interrupt control and can be designated to be controlled by any communication user key panel. This is where the producer pushes a "call button" on their user panel, which activates the circuit to talk to the announcers. The producer can talk privately to the announcer that is assigned to that specific call button or talk to all the IFBs. Normally, IFB 1 is the booth play-by-play announcer and IFB 2 is the color announcer. When the producer pushes the IFB 1 button, he will talk directly in the headset on the play-by-play announcer.

There are still dedicated IFB interrupt panels in use because some of the Telex designs and components have upward compatibility. The Telex system has been prevalent in broadcasting in North America for over 20 years and is a very modular and scaleable design. The IFB power supply creates either four or eight discrete IFB channels and input sources are programmed through the communications matrix. Additional audio functions, like a feed directly to a public address system, are easily accomplished in a matrix system.

**Figure 4.8** Sixteen-channel communications base station with an additional IFB control panel above it.

# PLs—Communications behind the Camera

In addition to the on-air talent working in the announce booth, other support and technical people are integral to organizing the broadcast. The system that all engineering and support personnel communicate over is generically called a PL. PL is an old telephone term that means "party line," stemming from the early days of telephones when several people would share a telephone line. There will be a stage manager, statistician, booth A2, and maybe a spotter, and for everyone to accomplish their job they must be able to clearly communicate with each other. In the announce booth, the stage manager will need to talk with the show producer, the booth statistician will need to talk to the event statistician, and the chief spotter will want to talk to the producer and any other spotters. As you add people to the production, you are adding complexity to the communications matrix and every communications circuit must be programmed.

In broadcasting, the PL system is either a point-to-point talk and listen or a party line system. Point-to-point communications is programmable so a communications channel can be established from any user to any user and provides private two-way communications. A point-to-point circuit is programmed in the communications matrix and is called a "port." A communications port can include as few as two users and as many users as desired, with a volume adjustment between users. A party line is a channel or line with a dedicated purpose. For example, there is a communications channel for the producer, director, scoring, tape operators, audio, engineering and up to 15 different channels or more with the newer systems. On each channel, you can have as many operators as required, so if the producer wanted to talk to everyone on a single channel, that can be accomplished as easily as assigning every operator to a separate channel.

Maintaining and installing the communications system is a critical job responsibility and requires good installation practices and troubleshooting skills (Figure 4.9). The A2 is the maintenance interface and needs to have patience with operators that are unskilled or unfamiliar with the equipment.

**Figure 4.9** This audio assistant is testing the circuit and talking to the truck.

In television, if a piece of equipment has a microphone, speaker or headset attached, it is classified as audio equipment and this includes PLs. In addition to PLs in the announce booth, there will also be a variety of communications required at other locations to accomplish the telecast.

The announce booth is the office or work space for the on-air presenters and may become home for several days. It is usually the responsibility of the audio assistant and the stage manager to maintain order in the house and keep the work space neat and organized. Often the announcers believe that they are "stars" and should be treated accordingly. This can sometimes make for some problems with the crew in the announce booth. For the booth audio assistant, a good bedside manner is as useful as any technical skills that may be needed.

## The Side-line Reporters

It has become common in sports broadcasting to have an announcer on the field or in the pits to get closer to the athletes and coaches. A system that consists of a microphone to talk on-air and off-air and an IFB with earpiece to hear program sound and information from the producer is used. The side-line reporter can use either a wireless or a wired system. As the name suggests, a wireless system consists of a microphone with a transmitter to send voice to the television truck and a headset receiver to pick up program audio and direction cues from the producer.

**Figure 4.10** Notice that the reporter is talking back off-air to the producer before this interview.

The microphone signal travels to the television truck and is assigned to the mixing desk and prefader level to a speaker for the producer to hear without going to the television program. Prefade means a full volume level before the mixer fader or show mix.

At an automobile race, the pit announcers are wireless because of the distances that must be covered and for safety. A wired system could be used at smaller or low-budget events or where radio reception is not good, such as in locker rooms under a stadium. If it is considered to be critical to the production, a wired position may be called for as a backup. Note: At every event I have worked with a wireless announcer, there has been at least one if not more backup hard-wired positions.

# Effects and Atmosphere Microphones

In addition to all the announce positions, the audio assistant is responsible for installing and securing all microphones on the cameras and microphones close to the sport. At a basic television show, microphones will be on the cameras, but the nature of the television production and the audio mixer's own experience will dictate the microphone selection and placement.

Microphone selection is often restricted by the quality and quantity that the television truck carries. A normal microphone complement on a medium-to-large television truck is ten shotguns, eight lavalieres and ten hand microphones. As you can tell from the inventory, there are no specialty microphones, so there are limitations in the microphone scheme. Specialty microphones have specific uses and are often not included on the television truck. Boundary microphones, contact and hydrophones are not normal equipment for a truck that is doing basketball or baseball half the year, but should be brought in when televising gymnastics or aquatics (see Figure 4.11).

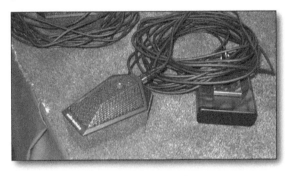

**Figures 4.11 (a) and (b)** The boundary microphone and miniature shotgun can be mounted very close to the sound source.

Microphones are often placed very close to the athlete and even on pieces of sporting equipment. For the audio coverage of the Olympics, there are over 20 different models of microphone from a single supplier. When you want to place microphones close to the athletes or on the field, they must be approved by a governing body for that sport. All sports are regulated by a sports federation specific to that sport. These sports federations have absolute jurisdiction in the placement of equipment on the "field of play."

**Figure 4.12** Large diaphragm microphone (AT4050) with Rycote windshield used for surround ambiance.

It has become common for specific microphone selection to be included in rental negotiations or to have microphones brought in as a separate package. Personal experience will influence microphone placement. This experience is gained by working in the field as an audio assist and sub-mixing the effects for other audio mixers. Many large productions will include a basic microphone layout, but the secrets of the trade are learned by networking with other sound people. There is a growing demand from ABC, NBC, CBS, FOX and ESPN for a higher quality sound in sports programming, and this is achieved with good microphones and good placement.

**Figure 4.13** Microphones on cameras, with a boom arm to keep the microphone away from the lens and keep it from interfering with the pick-up of the microphone. Do not lay them flat on the top of the camera lens.

Proper installation is critical to the accurate operation of this equipment. Figure 4.13 shows a side mount that puts the microphone above and away from the camera lens. As the camera operator tracks the action, the microphone is following the action as well. (See Chapter 7 for an in-depth examination of microphones.)

Sports television production tends to be a well-documented process, because most major sporting events are covered by more than a single broadcaster. Additionally all sports have the jurisdiction of a sports federation or sanctioning body, as mentioned previously, which has a significant amount of control over positioning of broadcasting equipment. Normally the production plan is submitted to the appropriate sports federation or governing body.

At international events the broadcasters are called "rights holders" and pay a fee to telecast the production in their home country or channel. In most cases, a production guide is prepared with a map of all camera and microphone positions, plus other essential information needed for the rights holders to produce the event.

Page 25 shows a very detailed map and page 26 is the mult-sheet that lists the microphones for the coverage of basketball. The audio assistant is responsible for setting up the microphones on the field of play, this map of the courts and the mult-sheet that lists the microphone position and model number are essential and must be followed in every detail. First, this microphone plan has been tested or is a standard type of configuration that meets the sound requirements of the sport, and, secondly, the sports federation has had to approve certain microphone placements, particularly the microphones that are attached to the badminton table. Generally, any deviation from these plans is unacceptable.

## Audio Troubleshooting

The most important skill for audio assistants is to be able to effectively and quickly resolve any problems with the audio systems that they are responsible for. Television field production by its very nature introduces additional variables that will inevitably interrupt the function of some equipment. Wet weather is a problem, given that many televised sports are played outdoors and will continue in the rain. Problems will arise and you will find yourself in the middle of trying to determine exactly why this piece of equipment is not functioning properly.

Electrical components have amazing reliability and television equipment is built to very durable specifications. You will find most problems to be in the interface and specifically in a piece of wire and/or connector. It will be one of your responsibilities to do fundamental maintenance and troubleshooting in the field.

Troubleshooting requires certain tools to facilitate the process. The basics include a voltmeter, battery powered listening speaker and a multipurpose tool. Most television devices are electrical and basic problem detection will require a way to detect and measure voltages. The device to the left in Figure 4.14 has two probes and is used to measure a variety of electrical properties.

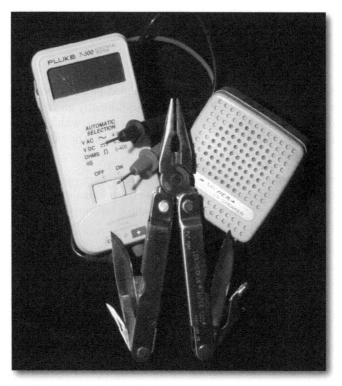

**Figure 4.14** The device with two probes is a voltmeter.

The device to the right in Figure 4.14 is a battery powered amplifier and speaker. Many audio circuits are analog and can be easily tested with an amplification device. Finally a multifunction tool such as a "Leatherman" is compact and extremely useful. This multitool device has become a standard in the broadcasting industry with its variety of screwdrivers, cutters, knives and pliers.

The number one rule for effective troubleshooting is to use a methodical approach that eliminates one potential cause at a time. First, check the obvious. Is everything plugged in correctly to the appropriate piece of equipment and patched to its destination?

The second thing is to replace the particular piece of equipment you have in the field, and if that does not solve the problem then it will be in the cables.

## Troubleshooting Microphone Cables

Next, we will examine the wire. Television audio uses two distinctly different types of cable: shielded and unshielded.

Microphones generate very small amounts of electrical voltage when they convert an acoustical sound to an electrical representation. From the output of the microphone forward, a series of amplifiers make the signal louder and usable. Additionally these amplifiers will make any defects or problems in the audio cable louder. Good-quality shielding around audio wires help to minimize any outside electrical or magnetic interference that may be in proximity to the microphone cable. The longer the cable, the more likelihood of a problem.

Now check the ends of the cable. A steady buzz or intermittent crackle usually indicates a problem around the connector. Connectors are either soldered on or screwed down on terminal blocks or binding posts. Solder joints are prone to become brittle with age and abuse, and are a constant source of aggravation to the field audio assistant.

Always test the cables before you leave the broadcast area. A simple box with three lights will tell you if there is a continuous circuit. However, this type of box will not tell you if it is a "noisy" cable. Many audio mixers will listen for noise through the mixing desk before anything is ever plugged in.

Be cautious of the path of any audio cable and avoid close proximity or running in parallel to bundles of camera cables, any high voltage cables, transformers or dimmer cables for lighting. When the audio mixer complains of a constant hum in the microphones, this is indicative of cables in close proximity to a high-voltage electrical cable. This will certainly induce a 60-hertz hum into your audio. Laying out cable seems to be a mundane task, but this is often where the problems begin.

Crackling when the cable is flexed back and forth indicates extremely weathered or old cable, or cheap cable. Do not use it.

Distortion is induced when the level of output signal for a piece of equipment is greater than the input capabilities of the receiving piece of equipment. This will occur when the microphone output is too strong for the microphone preamplifier.

During the Olympics many microphones are placed at a great distance away from the mixing desk. A device called a microphone-to-line amplifier is often used to amplify the weaker microphone signal close to the microphone source. The input off the microphone preamplifier will have an attenuation switch with a value of –20 dB. This attenuation switch reduces the sensitivity of the input of the device and will generally eliminate any overloading and distortion. Note: Audio Technica microphones tend to have a very strong output.

The microphone preamplifier is essential when using unshielded wire. This in-line microphone preamplifier will eliminate any potential buzzes or hums in long distances of shielded cables.

## Troubleshooting Communications Cables

While microphone signals flow to the mobile truck at very low voltages, communications are accomplished through another set of wires with voltage and the signal flow going away from the truck. To have communications from production and engineering to field operators, a headset and communications box is plugged into a cable that runs to the mobile truck. This cable is plugged into a communications channel (PL) so the operators can talk to the truck.

This remote communications device requires voltage on the wire to "power-up." This is a DC electrical circuit with a positive and negative voltages. This polarity cannot be swapped.

Communications systems are generally very reliable but can be prone to problems because of connectivity. Two wire systems were very popular in North America because of communications coverage needed for large sports like golf and car racing.

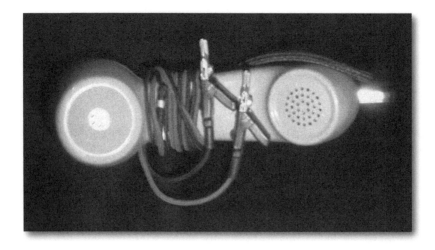

**Figure 4.15** This photo shows an analog telephone handset that was the main troubleshooting tool of the telephone company for decades. The new handset models perform the same functions but are more versatile in the digital world.

Many features and the functional operation of television PLs were based on telephone technology and the handset is a very useful tool. When the audio assistant arrives at a PL position, they can clip the handset to the wires to determine if there is voltage and if they are able to talk to the television truck. If the wire (circuit) passes these tests, then the audio assistant will install a television PL and retest the system.

Organize your cables so that the signal flow is in the direction of the male connector. This is usually the way the equipment that you must attach to the wire is designed; however, there are different schools of thought internationally. Microphone signals flow in the direction of the truck or camera. Communication signals flow away from the truck. After a thorough engineering facilities check, also known as a fax, the equipment is turned over to production to execute the telecast.

There is one very important rule when troubleshooting fiber cables: Never look into a fiber optics cable to see if there is light!

## Other Issues

Several other issues are important in audio troubleshooting, and these are discussed in the following paragraphs.

*Wind Noise.* A common problem with outdoor sports is wind. Wind blowing across the diaphragm of a microphone causes a fluttering, low-frequency sound that is heard as periodic rumbling distortion in the sound program. Foam and specially designed encasements are used throughout the Olympics. Always use the windshields and be careful, because they are fragile and expensive to replace.

*Isolation/Insulation of each microphone.* A microphone generates minute amounts of alternating current (AC) electricity. AC has the same characteristics as common house current. If you grab an electrical wire that has fallen on the ground, you will be electrocuted because you complete the natural electrical path to ground. However, birds are not electrocuted when they sit on one wire because there is no completed circuit or path to ground. This is why microphones are so prone to ground hums. If the microphone casing or any connectors can complete a path to ground, a hum will be induced. You cannot let a microphone touch the metal of a fence or mounting post because a ground may occur between the microphone and earth and a ground hum can be induced. Use a plastic Microphone clip or place electrical tape between any metal parts to insure isolation.

*Waterproofing.* You cannot leave audio connectors on the ground or exposed to weather because a ground hum can be induced into the sound when the metal jacket or moisture can conduct to earth. Always wrap connectors with a plastic sleeve that is taped along the jacket of the cable. The jacket connection to the sleeve should be seamless and without gaps to insure no water can trickle down.

**Figure 4.16** Here the audio assistant is at a secondary or sub-mix position. He has the responsibility to set up the equipment and operate it during the telecast.

# Summary

If you have looked at all the figures, you will have noticed that the audio assistant spends a lot of time in the elements. To quote the old Boy Scout phrase, "Be Prepared"! It is essential that the audio assistant have the proper tools, clothing and attitude. This is expected and an audio assistant will be judged by these criteria. When working an outdoor event like football, be prepared for bad weather. They do play football in the rain and it is the audio assistant's job to keep the equipment operational in all situations. I have done a football game as a hurricane approached the coast of Georgia, and the telecast stayed on air till the very end!

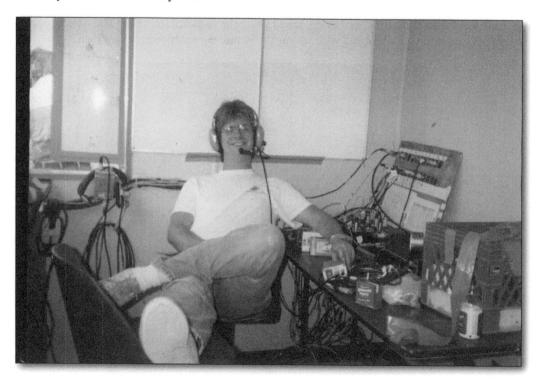

**Figure 4.17** The audio assistant should always be available by radio or headset. The wooden board against the wall was a row of female XLR connectors and a row of male XLR connectors that was soldered to a 25-pair bundle of unshielded telephone cable. The booth board was used for several years with virtually no failures. However, it was cumbersome to troubleshoot because the wires were hard-soldered and if a single wire was bad, you had to go to another connector and not just change out a single wire.

The sub-mixer position is the next step upward for an audio assistant and is where the assistant will learn critical listening skills.

# 5 Communication Systems

It is a lonely and frustrating feeling for the camera operator when the communication line is dead! Clear, concise and accurate transfer of information and commands between the production and technical crews is essential to a successful broadcast operation. Communication issues have been the demise of many good audio people!

At most sporting events, intercoms are the responsibility of the audio department. However, large entertainment shows and major events like the Olympics will have an entire department that takes care of the installation, interface and programming of the communications plan. Events like ESPN Extreme Games use multiple remote sites with telephone T1 lines and VOIP for communications between the production and technical crew at the remote operations.

The communications plan will define the users, equipment, features and size of a system, plus help determine the complexity and cost of a design. The category of production, music, sports or entertainment, along with the size of the production, are the first considerations for a communications systems design. The number of users and quantity of private and simultaneous conversations that must be supported will indicate how many channels of communications will be required. For example, most television operations will have the cameras and director share a "talk group" or separate channel of communication. This is the simplest means of communication, known as a party line and often referred to as a PL. The advantages of a PL include simple operation, easy expansion for additional user stations, and moderate costs. The disadvantages of a conference line PL system is there is no privacy between operators and lengthy conversations can be distracting when specific instructions are necessary.

Most original intercom designs were a single-channel party line shared by the director, camera operators, audio, lighting, stage managers and others. Contemporary party-line communications have evolved to a very sophisticated level that is robust enough for most applications.

## Powered Party Lines

The name "party line" came from the original telephone term where everyone shared the same line and could hear all conversations at once. In the early 1970s, Clear-Com built a commercial party-line system that became well-established in the concert and stage industry, but another company was soon to establish the television standards. Most original television intercom systems used a power supply to establish a bidirectional communication channel just like a telephone. Refer to Figure 5.1 for a diagram of a party-line system.

### Clear-Com Party-Line Intercom

Generic single channel **Clear-Com®** system.

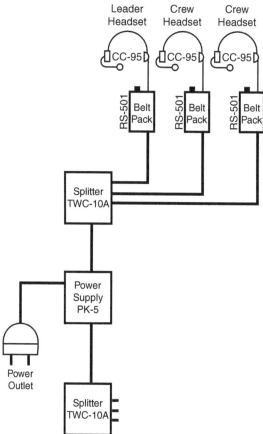

**Figure 5.1** Party-line system diagram.

In a basic party-line design, all of the users talk and listen to each other as a group. A system consists of beltpacks and/or rack-mount intercom stations and all of the user stations power off a common power supply and are all connected together by a two-conductor shielded microphone cable. The intercom stations are powered by a DC electrical current flowing through the cabling. Each station contains a microphone amplifier and a power amplifier for the headset or speaker and a circuit, known as a "hybrid," that allows the two-way audio signal to flow through the cable.

Additional channels were needed so that crew members could go to their own channel to communicate. When more communication channels were needed, additional power supplies were commissioned but there were issues with tying powered channels together. Most of these early systems were made by the television station's engineering department and were often based on telephone gear and technology. This approach lacked system compatibility and flexibility, plus expansion was often impossible beyond the original design.

I worked on several CBS mobile units in the early 1980s in which there were several different communication systems comixed. A three-channel powered party line provided a channel for camera conference, a channel for the producer and a shared engineering channel (ROH Systems, Atlanta, Georgia). There was an early McCurdy system, which was used for truck-to-truck and compound communications. Communications outside the compound were accomplished with basic telephone technology. For example, the spotter communications channel was a powered talk circuit common to all spotters and a chief spotter. The chief spotter had a modified headset to tap into the spotter party line and then switch to talk on the production intercom inside the TV truck. The show producer had a small mixer to balance the levels of the different people he needed to listen to and often spoke on separate microphones.

The obvious problem is that there was no integration of the various communication devices and it was necessary to modify headsets with switching boxes and summing amplifiers. The lack of integration was prone to electrical and operation problems and spurred the development of inter-communication systems. RTS was one of the first companies to take the existing technology of powered party lines and integrate a flexible package of user stations.

RTS designed a line of communication devices for the television industry that was modular, scaleable and portable. The RTS design is an integrated system with three basic blocks. At the heart of the architecture are self-powered, multichannel base stations that can talk to other self-powered base stations. The base station interfaces to an external powered talk network of beltpack users and to a program listen channel for talent, which can be interrupted. The row of keypads pictured in Figure 5.2 is the director's interrupt key for the IFB system.

It was the first all-in-one system where a producer or director could sit down with one headset and talk to everyone in the entire production team. For engineering, it was the first modular and scaleable system that was straightforward, flexible and easy to maintain.

**Figure 5.2** The director views the individual pictures and communicates by headset to the cameras and technical crew. The communications interface (pictured over his shoulder) is a 12-channel party-line system which allows the director to talk and listen on any or all of the channels. Notice that Director Bill McCoy is talking and listening on channel one, which is the director's channel and listening to channel two which is the producer's channel.

The master base station is generally in a fixed position like a mobile truck or TV studio, and the remote station is usually a stand-alone beltpack or similar circuitry integrated into a camera.

**Figure 5.3** This camera operator is using a tight fitting, noise-reducing headset that reduces the level of outside noise. I expect companies to introduce "noise canceling" headset very soon.

A master control station can access any or all of the PL channels within the communications grid. The communications grid is organized according to the production and a single channel could have as few as one person talking to another person, which is known as a private line, or more than two people, being a party line. The original 801 base station had six channels while the second generation 802 system averaged 12 channels; however, a station could be ordered with a maximum of 15 channels. The modular design of the RTS systems made it easy to add channels of powered communications and expand the system.

The 802 base stations could be manually configured to support various sizes of installation. The first six buttons of this station were generally configured for six channels of intercom and the last six sets of buttons could be configured as IFB interrupts or listen channels (Figures 5.4 and 5.5).

**Figure 5.4** This 802 could be just as easily wired and configured to be a 12-channel intercom system with the addition of two power supplies.

**Figure 5.5** The volume controls appeared on the front panel on the original 801 and were moved to an inside panel on the 802 as pictured. The geared knob is the volume control and the item below it is a rotary potentiometer to null the channel. The four rotaries on the bottom are microphone volume control, chime level and more.

Digital systems offer a variety of user stations and control panels that can be programmed to fit the needs of the operators.

The original RTS base stations were locally powered and used a multipair telephone cable to loop all of the base stations together. Each communication channel was discrete and hard-wired from station to station and each channel in the base stations used a discrete pair of wires to talk and listen over. For example, a 12-channel system used 12 pairs of wire to talk and listen and the interface was simple with copper teleco 25-pair wire.

The base stations terminate to a power grid that is the foundation for all external communications. The RTS system uses a power supply that generates DC power for three separate channels of voltage-driven talk and listen circuits similar to a telephone. For a large communications installation, additional power supplies (RTS PS 31) could be configured to give more channels. Additionally, each station has access to any of the discrete channels in the communication grid.

RTS Intercom Systems is strictly analog-based technology where a dedicated connection is made between parties.

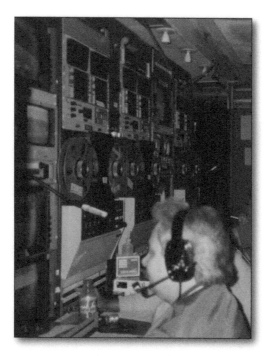

**Figure 5.6** Most technical positions and operators in the OB van wear headsets. This is a vintage photo with one-inch tape machines.

The external interface is usually a beltpack and headset that connects to the power supply and intercom line down a wire that originates at the side of the television truck. Each beltpack can receive two channels of communication down a single three-conductor XLR cable and is assignable to any two channels of the 12-channel matrix through an assignable switching wheel in the audio control room.

The RTS system is an assignable party-line system, where each communication channel can be private between individuals or party lines between groups of people.

## Two Wire, Four Wire and Telephone Hybrids

*The two-wire approach to communications is where a single pair of wires is used to talk and listen using a two-wire to four-wire converter, commonly known as a telephone hybrid. The telephone hybrid blocks the telephone voltage and matches the impedance of the telephone so the talk and listen circuits can interface with regular audio circuits. Digital hybrids use a special unit with DSP audio data compression and decompression at each end delivering audio bandwidths up to 15 kHz.*

*Telephone hybrids are used to send audio signals over phone lines, to integrate caller's voices into a broadcast. Audio signals over telephone lines are often used to send a back-up audio mix to the network and to send a "mix-minus" audio signal from the network to the remote operation. A "mix-minus" is the broadcast mix without (minus) the remote broadcast sound mix. The network mix-minus is how the studio announcers respond and interact with the remote announcers.*

*The problem with two-wire to four-wire converters is that they tend to degrade the audio and nulling the talk/listen circuit can be difficult. There are disadvantages to a two-wire system because some problem situations affect the intelligibility of speech. External channels also have distance limitations because of voltage loss. I have seen some clever ways that audio assistants have come up with to deliver communications to a remote camera location.*

*The alternative to a two-wire talk/listen is a four-wire design that uses a pair of wires for the talk circuit and a separate pair of wires for the listen circuit. The greatest advantage to a four-wire system is that speech intelligibility is improved because now you are using higher fidelity audio signals and paths.*

The advantages of party-line systems are that they provide clear and reliable communications, they are portable, relatively easy to set up, use standard microphone or telephone cable, and are simple to operate. The typical requirements of an intercom system often aren't that great and a conventional party-line system is often sufficient. Most users only converse on one or two channels, receiving cues from the directors and talking within their own group. The directors talk with the various groups (channels) individually or all together, and there may be an additional channel for directors and producers only.

Party-line intercoms such as Clear-Com are standard equipment at performing facilities, community theaters, and school and college auditoriums, allowing communications among the lighting and stage-production staff. Touring sound-and-lighting companies will carry these systems with them for concerts and events.

Smaller TV broadcast facilities and mobile production vehicles will have up to six channels of RTS party-line systems to communicate among camera operators, audio and video technicians, outside broadcast staff, facility technicians and the talent. Two-wire party-line systems are still common but as the communications needs become greater and more complex, a point-to-point matrix intercom system should be considered.

## Matrix Intercom Systems

A matrix intercom is a point-to-point communication system where user stations are connected to a central communications mainframe through which all audio and control data are routed. Point-to-point communications allows the users to call a specific station for a private conversation or to call a "talk group" such as the camera conference to give directions to all cameras assigned to that group. Matrix communications systems are generally computer programmable and feature tally lights at the receiver stations indicating the caller ID. The original matrix intercom systems were analog based with dedicated wire and control to each station. Contemporary systems are digital with audio signals and control converted to digital signals and routed through fiber, coax and even CAT5.

Programmable point-to-point matrices were the predecessors to modern intercom systems and McCurdy Radio of Canada built some of the first matrix intercom systems. In the 1950s McCurdy built the first matrix intercom system, in which a relay closed a cross point between wires making a connection. Relays are large remote-controlled switches that require a lot of power. With the introduction of solid-state technology, relays were replaced with solid-state transistors, which increased capacity and reduced the size of the system. However, it still was a hard-wired cross point matrix that requires a dedicated wire to talk and one to listen. A matrix design is governed by a mathematical square-law relationship. A five-user matrix requires 25 cross points. To double the number of users to ten, the number of cross points is squared for a total of 100. (See Figure 5.7.)

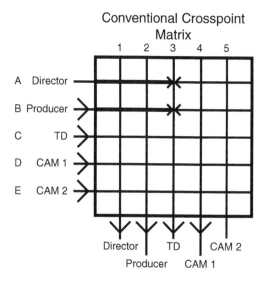

**Figure 5.7** A true matrix system has a fixed number of inputs by a fixed number of outputs and a communication channel is established by closing a switch. The problem with this design was that each talk path required switch circuit plus a dedicated talk wire and listen wire and this design is soon hampered by the size.

In the late 1970s microprocessors were available and in the 1980s McCurdy built the first intelligent intercom system where data was sent to the user station. NBC commissioned a system for the 1988 Seoul Olympics that was 350 ports and required ten full racks, 20 kW of power and weighed two tons. This was the largest matrix intercom system ever built and provided many advanced feature, but this engineering design was soon to be replaced with a digital solution. To increase the flexibility and to eliminate the need for a large number of interface wires, a digital system was devised. Once again borrowing technology from the telephone industry, this method continuously sends bursts of data to the user station that is embedded with audio and data to reconstitute the audio information.

Digital systems are fully programmable so that any type of point-to-point, group or party-line communication can be created, and any or all of these interfaces can be accessed by each user station. A digital matrix system offers a variety of multikey intercom stations that allow multiple channels of communication. The stations themselves typically offer between 4 and 32 keys each, and are often expandable.

Digital matrix systems also allow the voice levels of the individual keys to be controlled at the stations, giving each user his or her own audio mix. No matter how many individual keys are used, the stations connect to the central communications frame via a single four-pair cable for analog transmission or single-pair or coaxial cable for digital transmission.

Digital systems have a high degree of flexibility and interface easily to other systems such as telephones, two-way radios, a party-line intercom, audio and IFB/cue systems, microwave, and ISDN links and relays to control other devices. The digital base station has many similarities in looks and functions to its predecessor, the 802 and 803. In the digital systems the buttons are programmed in each station.

## Time Division Multiplexing (TDM)

By the early 1990s, intercom systems were beginning to utilize another process pioneered by the telephone industry. Time division multiplexing (TDM) is a method where two or more signals are transmitted on the same path, but at different times. This type of multiplexing combines different data streams by assigning each stream a different digital time slot in a set of slots. A digital time slot is reserved for each data stream and TDM repeatedly transmits a fixed sequence of time slots over a single transmission channel. In a voice conversation, the time slot is reserved even though 50% of the time only one person is talking. (See Figure 5.8.)

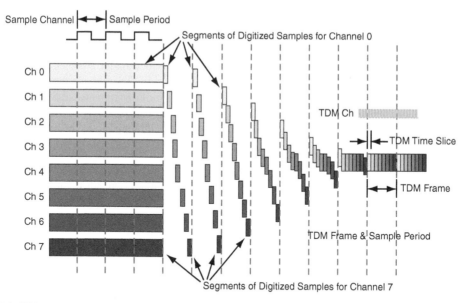

**Figure 5.8** TDM data packs.

RTS and other systems create digital streams for each conversation (audio signal) by using a process called pulse code modulation (PCM) and apply the digital conversations to the TDM data stream. Each channel of communications is continuously sampled into the TDM data stream.

The big difference with conventional cross point communications and the TDM-based system is that the cross point method is either on or off and the TDM technique is capable of mixing the volume of the assigned channels. With TDM technology, modern intercom systems are capable of interconnecting any combination of user stations and even channels within a user station through software control. This ease of configuration provides different types of communication without any change of hardware. Additionally, a programmable system can easily configure a variety of functions such as a conference line with multiple users, an isolated line with two users, an IFB, a call group along with external communications over telephones and wireless. In addition to communication functions, ancillary functions such as tally lights, status lights, switch closures and more are available with a digital system.

The RTS system is completely modular and configures a group of eight communication channels or ports on a "port card" under the control of a single TDM data stream. Each port controls a different user panel and is completely programmable for a complex multichannel producer station or a simple IFB listen station for talent.

Port cards are harnessed together in a mainframe with other port cards and power supplies to configure a system with up to 1,000 different communications ports. Each port on a port card has an input, an output and a bidirectional data signal. The talk signal is carried on a pair of balanced wires separate from the listen signal, like a four-wire system, and the data is on its own pair of wires. Remember, unlike the four-wire and two-wire systems, a TDM digital system does not use voltage on the talk/listen lines.

Because the signals to and from the user stations are +8 dBu audio and RS-485 data, the user stations can be interconnected to the mainframe with copper wire, fiber optics, coax or even transmitted. The interface connectors on the mainframe of the RTS ADAM system are either RJ-12 phone connectors or a multipin DE-9 connector. For a nontraveling permanent installation, the plastic RJ-12 phone connector is okay, but for a remote truck it will be more reliable with a multipin connector. Each port will have an output connector breaking out to three balanced pairs of wire that can be terminated to an input/output panel on a remote truck or adapted into any carrier such as copper wire or fiber optics.

The port card has eight ports that share a single data line. The system is continuously sending a data stream to the user stations, polling each station one after another till all eight stations have been checked for any changes. The period to poll all eight stations is called a sample period, or a TDM frame. The polling period for each user station is usually less than 10 milliseconds.

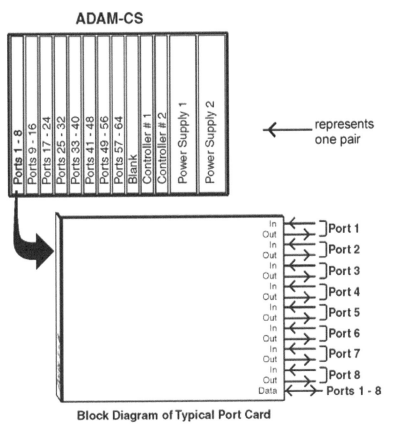

**Figure 5.9** The ADAM communication system uses cards with eight channels or ports of communications.

RTS makes two systems that have fixed numbers of ports and are not expandable, and a mainframe that is completely modular. The RTS ADAM system has a frame size that can accommodate between 8 and 136 ports in groups of eight. The ADAM system can operate multiple frames to build up to a thousand-port system.

The RTS system has the flexibility to couple two or more systems together through a method called *trunking*. Trunking is where two or more separate communication matrices act like user stations and can be accessed by each other. The trunked channel is a party line to those users assigned to it. RTS has an intelligent trunking system that assigns addresses to the matrices and can poll each port.

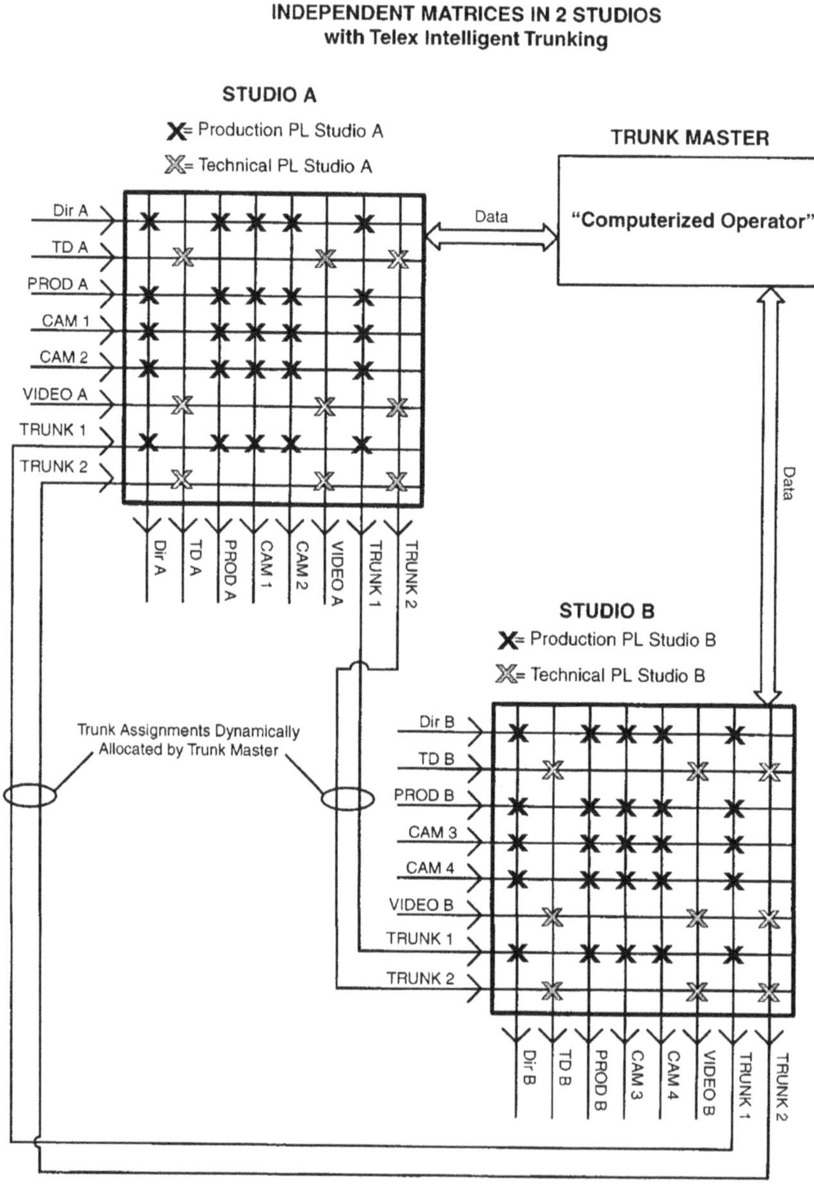

**INDEPENDENT MATRICES IN 2 STUDIOS**
**with Telex Intelligent Trunking**

**Figure 5.10** Two separate independent communication matrices can be connected together using a method called "trunking."

The need for sophisticated intercom systems brought several international manufacturers on to the scene with different solutions. Riedel Communications produces a system with essentially the same features as RTS that is dominant in the European markets. Digital matrix intercom systems are ideal for complex communications applications because they allow many channels and types of communication to be combined into a single, integrated system. These systems are flexible and programmable and any real-time signal routing changes can be made without moving any cables. Sophisticated interfacing capabilities make digital matrix intercom systems ideal for interfacing over telephone lines, cell phone or two-way radio.

Digital matrix systems can route multiple sources of program audio, create as many cue channels as desired, and control external devices such as lights and monitor levels via relays. Digital matrix intercom systems are the choice of most television broadcast studios and larger mobile productions. A digital matrix system does not completely eliminate the use of powered two-wire stations. User stations that need to be somewhat mobile but still can be wired are beltpacks (BP) and are generally two-wire units so they can operate on a single XLR cable. These stations are generally worn by the user who is away from the remote truck in the announce booth or the scoring center, usually somewhere away from the TV compound.

## Communications Using Voice-Over-Internet Protocol (VoIP)

Remote communications over long distances are required for television productions. They can be as simple as telephone communications from the television truck to master control, or as complex as IFB control from a control room in New York to locations anywhere in the world. At the 2006 Olympics, NBC produced curling in Torino from a remote studio in New Jersey using voice communications over data lines. At the ESPN X Games communications were handled from three different locations with camera, stage manager, lighting and audio PLs over common telephone T1 lines.

Voice over IP means sending voice information in digital packets of data over the internet rather than using the conventional telephone network. Equipment designers have embraced the VoIP protocol and technology and have integrated this into their systems. Telex® has designed the RVON-C interface card that converts analog audio to digital VoIP audio. The card(s) are installed directly into the ADAM Intercom frame; the RVON-8 provides voice-over-IP communications supporting eight audio channels (ports) in and out, plus data.

The RVON-8 card uses standard Ethernet protocols and is compatible with 10 BASE-T or 100 BASE-TX Ethernet compliant devices and networks. It provides a single RJ-45 Ethernet connection for use with a 10BASE-T or 100BASE-TX network. In order to be IP compliant, all cards have a unique MAC ID when shipped from the manufacturer. Typically, the MAC ID of a piece of hardware such as the RVON-8 card has a fixed or static address.

Each channel has configurable network and bandwidth parameters that can be tailored to an individual network. Bandwidth parameters are adjusted by an audio encoder that controls the quality of the audio you hear and the network bandwidth used. An audio encoder/decoder is known as a *Codec*. It uses an electrical and mathematical process called an *algorithm* to compress audio. The codec rate is the actual bits per second of the audio in compressed form. This is sent over the network through data packets. The packet size determines how much audio data is carried across the network in each transmitted packet.

The Codec type and packet size require different amounts of bandwidth from the network. The packet size you choose for the audio transfer will affect the audio you hear and the bandwidth you use over the network. The larger the audio packet you choose to use, the lower the bandwidth used. The larger packet size can result in a higher delay and longer gaps if the packet is lost. On the other hand, smaller packet sizes result in larger bandwidth use but lower delays and smaller gaps if the packet is lost. It is important to note that you must configure the channel settings on each end of a connection and ensure the same codec and packet size are selected at each end. Voice activity detection (VAD), when enabled and only when audio is above a certain threshold, sends packets. Otherwise, a silence packet is sent once, and not again until audio is above the threshold. If there is ever a need to have all audio paths open and active, a network designer must account for this scenario.

The RVON-8 card is mounted directly in the ADAM mainframe and is hot-swappable so if a card has to be added or exchanged you do not have to reboot the system. Telex uses a common software and the RVON-8 features are configurable through Telex's AZedit software. RVON-8 supports Telex® Intelligent Trunking over IP. (Trunking is a method of using relatively few audio paths for a large number of potential users.)

## System Design

The first and foremost consideration of a communications system is the needs of the operators. Communications is a function that facilitates the coordination and execution of the efforts of the team members and technicians in a synchronous and organized process. For example, television is a team activity that requires the coordinated efforts of on-camera talent, stage managers, directors, producers, camera operators, audio technicians, lighting, technical support plus staging, makeup and hair! If talent is not ready with makeup and hair, there is no reason to be recording! For example, the stage manager is the eyes, ears and brain of the production on the stage and is the interface with the various departments to insure things are progressing. This person needs a two-channel wireless intercom with interrupt into the house PA system.

Almost every broadcast production has a stage manager but the communications requirement for a stage production is completely different from a sports production. The stage manager for stage will require a wireless system because of the large area that they are responsible for, where the sports stage manager is usually closer to the talent and is usually wired to the intercom system.

The number of simultaneous and private conversations that are required will be key in determining how many and what kind of communication channels are needed. Since one of the principal functions of the director is to organize and cut between the cameras, it is customary to have a single party line for the director to talk to all the cameras. The original RTS systems had a special operator station for the video controller in the OB van to isolate the camera operators from the general camera conference. The director did not have any private conversation features unless an additional channel was established.

The advantage of a matrix intercom system is that a camera conference as well as private channels can be easily programmed. Programming flexibility is critical for outside broadcast because the television truck will be required to reconfigure on virtually every set-up.

### *Good Design—The Modern Announce Booth*

*In the modern announce booth, many events use fiber interconnect, which has unique set-up procedures. The NASCAR audio team uses a roll-around rack that houses fiber multiplexors, power supplies for intercom and IFB channels, preamplifiers for microphones and a talkback return. In the booth the PLs and IFB are powered, thus reducing long two-wire runs and the probability of hum and interference.*

## User Stations

User control panels range from a simple two-wire beltpack to multikey stations in the television truck. User stations are panel mount, stand-alone and come with a fixed number of assignable keys that can be programmed to be communications ports such as point-to-point, ISO-channel, party line (PL) or an IFB or relay function. User stations come with talk keys and separate listen keys. User stations either connect directly to the matrix with twisted pairs of wire or through four-wire to two-wire converters for powered beltpacks and IFB stations. The stations themselves typically offer between 4 and 32 keys each, and are often expandable. No matter how many individual keys are used, the stations connect to the central communications frame via a single four-pair cable for analog transmission or single-pair or coaxial cable for digital transmission. A very common broadcast user station is the BP-325.

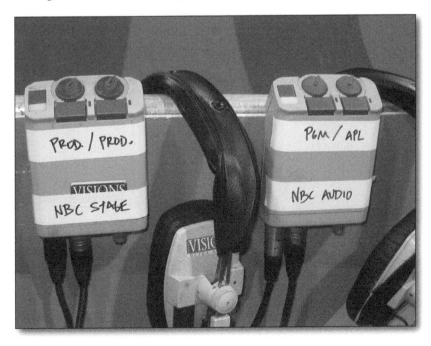

**Figure 5.11** This is a powered, two-wire, two-channel talk and listen user station.

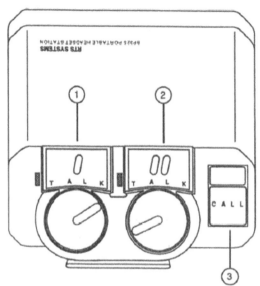

**Figures 5.12** Diagram of top view.

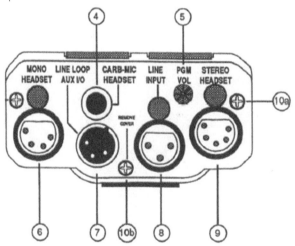

**Figures 5.13** Diagram of bottom view.

1. Channel 1 talk button, indicator light and listen volume control.
2. Channel 2 talk button, indicator light and listen volume control.
3. Call button and indicator light.
4. Carbon microphone headset jack.
5. Monaural headset jack with dynamic microphone.
6. Stereo headset jack with dynamic microphone.
7. Auxiliary volume control.
8. Intercom input connector.
9. Intercom loop through.
10. Housing screw.

The BP 325 can be programmed with internal dip-switches to lock on or lock out specific functions, such as microphone latch or call tallies.

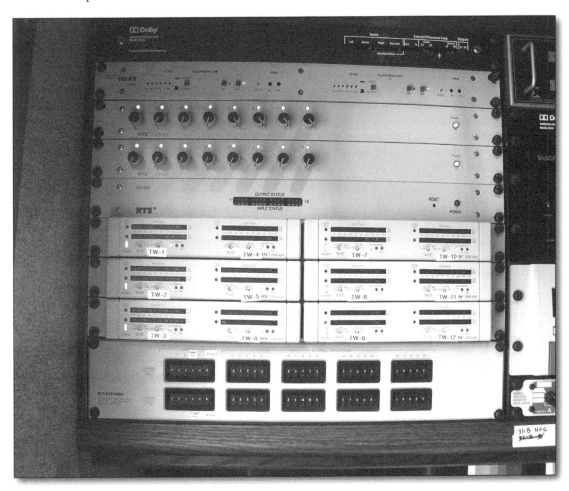

**Figures 5.14** SSA-424 digital auto-nulling two-wire to four-wire hybrid provides two independent channels of two-wire (TW) to four-wire hybrid. The TW interface is suitable for balanced and unbalanced systems, allowing RTS matrix systems to interface to virtually any party-line intercom system in use today. Easy-to-read front-panel level meters make matching audio levels simple, and the RTS advanced DSP hybrid eliminates all need for manual nulling, even under varying loads. A call signal option is available, which encodes and decodes TW call signals that can be used to interface matrix GPI circuits.

Every operator must have a user station to communicate within the system. The number of channels and features are scalable to the needs of the operator. For example, a playback operator will need fewer channels of communication than the producer and a vision operator will need direct point-to-point contact with the cameras. User stations in the OB van are mounted in the operation areas often in standard 19-inch rack widths and may have from 4 to 32 programmable channels of communications. The stations have an ID name window that flashes when a caller is talking.

Stand-alone desktop stations are useful during the setup because they usually have a speaker that can be left on for monitoring. There should be some common sense when using intercom systems especially with open microphones and speakers. This has been the demise of a couple of television people!

Headsets, speakers and microphones are the human interface to the communications systems. The proper selection of headsets can make a tremendous difference in the operation of the communications system. Single-ear headsets are comfortable but are not as practical as double ear headsets in high-noise environments. Dual-ear headsets also allow intercom channels and program signals to be split between the left or right ear, which helps separate the signals in complex listening situations. The audio mixer usually does not wear a headset and needs a listening station with a speaker and microphone.

The quality of a headset is not only indicative of the durability and sound quality, but also the comfort and operator preferences. Cheap components such as ear and microphone transducers will impact the speech intelligibility and compromise the performance of a communications system. Some operators will prefer to use their own equipment, which can be problematic to a system because of the range of impedances and sensitivities of equipment.

## Interruptible Feedback (IFB)

The final critical link in the communications system is the interface with the on-camera talent. The device that allows an announcer to hear the program sound and be talked to by a producer is called an IFB or interruptible foldback or feedback. The IFB is a listen-only device with an audio feed and a one-way intercom. Normally a mono mix is sent to the announcer so that each ear hears the same mix; however, only one side of the announcer's headset will interrupt with the producer's voice and with the mono mix the announcer hears everything in the mix.

The term IFB is generic because there are many engineering methods used to accomplish the task. For example, a radio with an earpiece can be an IFB just like a beltpack and headset is an IFB. I experienced a CBS method that had amplifiers in the television truck and the audio person controlled the volume levels to the announcers from the truck. This design offers better quality sound to the announcer, but it burdened the audio person with controlling the announcer's headset levels because the operator had no controls over this primitive system.

The RTS brand of IFB system is a powered two-wire system for mobility and flexibility. The original IFB 4000 offered a mainframe that powered and controlled up to 12 separate and discrete IFB listen channels with unbalanced audio and DC operating power to IFB beltpacks. The IFB remote box is cabled with three wires back to the mainframe and its powered electronics deliver two channels of listen to each headset.

The IFB mainframe is located in the audio room where the inputs and outputs will reside in the patch bay. The IFB mainframe has three separate stereo inputs that can be assigned to any or all of the IFB channels. (See Figure 5.15.) Each IFB channel can be assigned to any one of three input sources that are best thought of as three stereo mixes and designated Input A Left/Right, Input B Left/Right and Input C Left/Right. At the IFB mainframe there are left and right volume controls for each channel of IFB. The mainframe volume control sets the nominal operating levels and the remote IFB pack is where the announcer can adjust their own levels as necessary.

A sports announcer needs a way to talk back to the producer without going out to air. This is accomplished by a switch that toggles between the announcer's microphone going to on-air or off-air to the producer. For many years this was handled by a passive momentary switch that had a normal output and a switched output that would feed to a speaker or headset. The announcer's microphone output begins at a low level and was prone to pops when switched. Amplifying the signal to line level was a common solution, but this added another device with a bundle of wires to the announce position and the entire mess was prone to problems. TV trucks began to construct an announce box that incorporated the IFB, talkback, amplifiers plus inputs and outputs into a package that was fast to install and easy to maintain.

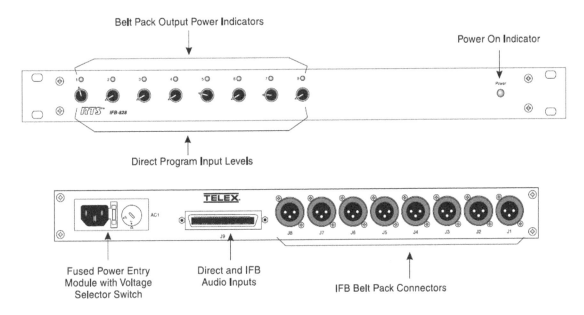

**Figure 5.15** The current IFB system is the IFB-828, which powers up to eight RTS Systems' Model IFB-325, 4020, or 4030 IFB beltpacks and interfaces to any RTS digital matrix or two-wire intercom system.

Some broadcast set-ups will locate the IFB-828 in the audio booth near the talent location and program the ADAM system. The IFB 828 can easily power beltpacks up to several hundred feet away, but then the beltpacks may result in diminished performance due to DC resistance in the cabling and noise induced by surrounding equipment. Use appropriate precautions, such as a heavier gauge, shielded cable, and always route audio away from unshielded equipment.

Any intercom port can be programmed to be an IFB by using the ADAM edit software and a computer. Select the IFB button on the ADAM edit toolbar and then double-click on the IFB to open the Edit IFB dialog window for that IFB. For each IFB that you set up, the intercom port must be set for a program input port in the input text box and the IFB output port must be set in the output text box.

## IFB Stations

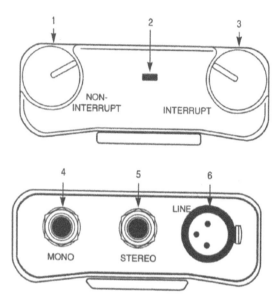

1. Channel volume control
2. Power indicator
3. Channel volume control
4. Mono headphone jack
5. Stereo headphone jack
6. Input connector

**Figures 5.16** The IFB user station has mono and stereo output with volume control for the channels. A common mistake with an inexperienced A2 is to quickly plug a stereo tip-ring-sleeve jack into the mono input! It will work, but not correctly.

The IFB can be used anywhere a program feed is needed, such as with performers with wireless in-ear receiver monitors. In-ear monitors have become an essential element in entertainment productions and can be programmed for an interrupt of the audio feed just like the IFB. The first major use of in-ear monitors was by intercom designer Larry Estrine at the Opening Ceremonies of the Sydney Olympics. There, 1,000 tap dancers were fed the music directly into their ears, which made it easier for them to dance in sync with the music disregarding the acoustical delay of the PA. Because a matrix system can mix signals, the director could communicate to the performers during the performance.

Matrix intercom systems began as a simple closure to activate a circuit. Most contemporary designs not only permit a circuit to be activated but to also control the volume level of the incoming and outgoing audio levels of each intercom and IFB channel.

The LCP-102 panel for matrix systems controls any one of three functions on any station. In TRIMS mode, it can be used to adjust any input or output audio levels for the matrix. In PAP mode, you can select any input source for an IFB and adjust the level of the source for each of the IFBs. In the CDP mode, you can make assignments to party lines and adjust listen levels for any party-line member. In all modes the LED displays provide full details on the action being performed.

### Matrix Audio Distribution

*The quality of matrix-based systems has advanced to where the matrix is the input/output of all audio signals for distribution. The systems have full bandwidth and offer very high-quality audio. Some systems have included DSP (digital signal processing) in the circuitry.*

# Wireless Intercom Systems

In most applications, the wireless intercom is an extension of the wired system, used by those who require mobility for safety or convenience. Because it is a complex technology that's difficult to manufacture well, wireless intercom systems are often considerably more expensive per user than wired.

So that users can talk together at the same time, each beltpack must operate on its own unique frequency—again, just like wireless microphones. Each beltpack has a receiver in the base station tuned to its transmit frequency, and the signals from the beltpacks are adjusted in level and placed on the audio bus of the base station. The mix of all of the beltpack conversation, plus that of any connected wired intercom or program source, is then retransmitted from the base station to receivers in all of the beltpacks, completing the loop. A six-up system with a base station and six beltpacks uses seven different frequencies.

Wireless intercom systems consist of full-duplex beltpacks (able to converse two ways) and a base station (Figure 5.17). Each wireless beltpack requires a transmitter and a receiver working together closely, plus audio circuitry, a mic preamp and a headphone amp. This fact makes them inherently more complex than wired beltpacks and more difficult to manufacture.

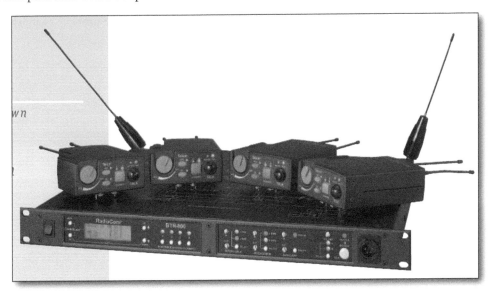

**Figure 5.17** Wireless intercom systems consist of full-duplex beltpacks (able to converse two ways) and a base station.

Before using wireless intercoms or microphones, the user must determine what frequencies are being used in the vicinity and what other wireless devices are used in the installation: microphones, two-way radios, walkie-talkies, and so on. Getting a complete picture of the radio frequency environment is the first step in frequency coordination so that the best and most interference-free frequencies may be chosen. This is typically done by using a computer program that calculates all potential interactions among the existing frequencies and the new ones that the wireless intercom will introduce. This process is essential when using crystal-based, single-frequency wireless intercom systems or microphones.

Frequency-agile wireless intercom systems allow the user to select different frequencies over a much wider band—typically equivalent to several U.S. television channels—with the same device. This agility allows the user to move away from interfering frequencies at the touch of a button or with a computer program, and can reduce preplanning. However, remember that changing a frequency will also change the RF environment and can potentially cause interference effects elsewhere in the system.

The great advantage of wireless intercom versus other two-way radio technologies is that it is full duplex, so that users can maintain a true conversation, hands-free instead of push-to-talk. Also, they are designed to work in conjunction with wired intercom systems, with the proper interfaces and connections to do so.

Wireless IFB systems are also typically used as additions to wired IFB systems. Wireless IFB systems are used for listen-only applications such as monitoring an ongoing program, cueing talent, and where the return path for the person's voice is a microphone (e.g., an on-camera television announcer receiving cues from a director). These units can have an additional feature that allows the incoming program material to be interrupted or ducked when a voice cue is introduced.

The audio from an intercom line, a program source or both is input into a transmitter. One or more beltpack receivers with headphones or ear buds are tuned to receive the signal from the transmitter. Some of these beltpacks have two receivers on different frequencies, so that the user can switch between two different audio sources.

# 6 Cable—Hooking It All Up

Wire is a generic commodity used in many different television applications such as microphone cable, communication PLs, video coax, electrical power, CCTV, and telephones. Copper cable is easy to use and requires no special encoding box but can be prone to problems because audio circuits require very low AC voltage. Fiber-optics cable has replaced wire cable in many applications because fiber can carry a large capacity of audio signals in a small size and is not susceptible to electromagnetic interference. Television requires a lot of connectivity that demands quality and reliability. No matter how superior the mixing console or microphone is, bad or cheap cabling will certainly cause signal degradation or failure.

Cable provides the connectivity to transfer electrical charges or light pulses from point to point and television uses both copper wire and fiber-optics cables. Copper wire is the preferred metal for analog audio because of its low resistance. Analog audio signals carried on copper wire only need to be amplified to be split, combined, summed, distributed or processed. Copper cable is also used in digital transport systems where wire coax and CAT 5 cable is used to transport the signals between encode-decode interface boxes. Copper wire is also used in cables for equipment using serial digital interfaces (SDI) cables.

Fiber optics is becoming the preferred choice for digital audio transport because of the small cable size, superior durability plus high audio capacity. Fiber and all digital transport solutions require the audio to be processed through analog-to-digital converters and digital-to-analog converters to restore the analog signal. Many digital-mixing consoles can accept fiber and coax inputs and outputs.

The television truck must maintain maximum flexibility to configure a system that can support a variety of different productions. Analog and digital interfaces are common to interconnect broadcast equipment like mixing consoles and recorders. Even though there is an ever-increasing amount of audio equipment that has both optical and digital inputs and outputs, copper wire is still needed to provide analog interface to the mixing console with the patchbay and to the I/O panel of the television truck. Copper mults are still used extensively in sports television for communications and on-air announce microphone and headset feeds to the television truck. These lengths tend to be less than 500 feet, so the microphone should not have to be remotely amplified.

Sports and entertainment productions are growing exponentially along with domestic and international distribution. Interconnectivity is critical for large-scale television productions like the Superbowl, Academy Awards and the Daytona 500 where multiple splits, mixes and audio feeds are required. These signals must be transferred, combined and distributed among many recipients in time and phase. This is accomplished over analog copper cables, digital mults and using digital language such as MADI directly into routers or mixing console.

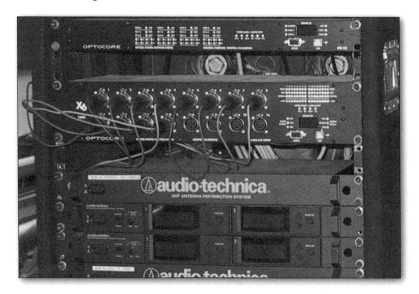

**Figure 6.1** This audio position uses a 16-channel "digital mult" to input the audio from a remote RF site.

A key requirement for copper and fiber mults is the cable's ability to withstand pulling strain and maintain flexibility. Copper wire will vary in diameter, gauge, construction, capacitance and resistance. Pure copper is an excellent conductor but lacks tensile strength, while aluminium is a strong metal but has too much resistance for audio circuits. Additionally, more strands of wire in a cable usually will yield a more flexible cable and reduce the possibilities of metal fatigue, thus increasing reliability and durability.

Copper cable used for audio will have shielding to minimize electrostatic noise. Any noise entering the cable will travel to ground through the shield and not interfere with the inner conductors. The shield density and shield construction, whether braided wrap or foil shield, affect performance. Metal foil provides effective shielding but may not be strong or flexible. A foil shield can break down if excessively flexed, while braided shields are more flexible and common for microphone cable. Additionally, magnetic fields from lighting ballast or electric motors are cancelled by common mode rejection in a balanced line.

**Figure 6.2** This is the "termination room" at the Jacksonville Stadium for the 2005 Superbowl.

# Good Audio Practices

With copper cables, every connection point is a potential problem area because of contact resistance and inductance between conductors. Here are some pointers:

1) Do not mix low-level microphone signals with voltage from PLs and IFBs.

2) Use balanced cable with two center conductors. Unbalanced cable has a single center conductor and uses the shield as a return path for the signal.

3) Always use strain relief and weatherproof audio connection points. Wet connectors will become a potential source of ground hums.

4) Maintain proper distance from any interfering source. High-voltage power cables will certainly inject interference into a copper audio mult.

## Analog Audio Signals

*Even though use of fiber cable is growing, there will still be a need to interface analog devices like microphones and processors. Microphones and professional audio equipment use a balanced audio circuit to transfer audio signals. A "balanced line" is an audio signal transmitted over a minimum of two wires with a third wire that is the ground or shield. When the balanced audio is received by a transformer or an active differential amplifier, only the difference between the two signals is detected as audio. Signals present on both wires, such as noise or interference, is cancelled.*

*On professional audio cables, the shield is never used for a return signal path as in unbalanced cables. It is only used to equalize grounds or commons, and to drain off noise and interference to keep it from infecting the audio signal. Unbalanced line is on a two-conductor circuit and one of the wires is the shield or is grounded.*

*The XLR-3 type of connector is used worldwide for the interconnection of professional audio equipment using balanced line principles. However, using an XLR connector in no way indicates whether an audio circuit is balanced or unbalanced, line or microphone level. It is the professional audio standard for audio outputs to be present on XLR male connectors and for audio inputs to be XLR female. The RCA plug is an unbalanced line-level connector and requires a transformer to connect an audio feed to balanced equipment.*

*With analog audio there are three level ranges you are most likely to deal with: −10 dB, +4 dB and microphone level. Microphone levels are usually somewhere in the range of 40–60 dB lower than line level and those signals need to be amplified before going to tape, or a mixer channel, or digital converters. Only microphones or other transducers will create a low-level analog audio signal and it is very desirable to amplify these signals as quickly as possible.*

*Line-level analog audio signals can be processed, split and combined at a line-level signal strength. Line levels range from −10 dB for unbalanced equipment to +4 dB or +8 dB line level for professional balanced equipment. Digital inputs should be measured where 0 dBFS (full-scale measurement) is maximum level and anything beyond will be unusable.*

## System Connectivity

The television truck is the center of all broadcast functions and operations and must be connected with the various pieces of electrical equipment through copper or fiber cables. For convenience and safety, bundles of individual copper or fiber pairs are packaged and bundled together. Bundles of pairs of copper or fiber cable packaged together are called mults which is short for "multi-pair." Analog mults will generally have a multipin connector that can be connected to a stage box fan out of XLR connectors or plugged directly into the input and outputs (I/O) of the television truck and accessed in the audio patchfield. Television trucks with a digital mixing console will have digital I/Os that route directly to the mixing desk.

**Figure 6.3** Ron Scalise and Mark Emmons review interconnect at a remote OB van that was used to sub-mix seven host positions for ESPN Summer 2005 X Games.

Connectivity with copper cable is still used in short runs with some OB van and compound interface. The individual pairs of wire for analog circuits use a globally standardized interface connector know as an XLR. The XLR uses 3-pin connectors to electrically join balance analog audio circuits. Usually 12 individually shielded wires are bundled in what is commonly known as a "mult." Mult is an abbreviation for the bundle of 12 wires that is enclosed in a weatherproof and sturdy jacket that is cut to manageable lengths—usually 50 or 100 meters.

Connection compatability is more of a problem in Europe than in North America. The 1984 Los Angeles Olympics standardized the durable multi-pin connector known as the DT-12 as the interface connector of choice in North America. A DT-12 connector has 36 individual pin connectors to electrically join 36 separate pieces of wire. Most North American OB vans have panel mounted DT-12 connectors that provide connectivity from the outside to the audio patchfield.

Optical fiber, coax and CAT5 cable is used to transport digital signals. The greatest advantage of digital mults is greater capacity in a smaller package. Fiber-optics cable utilizes a variety of connectors for inside and outside connectivity. Outside cable runs use heavy-duty military grade connectors and cable with Kevlar coating that prevents the cables from being ripped or damaged. Armored fiber-optic cable is approximately 5% of the weight and bulk of conventional multipair audio cables; this not only offers faster cabling turnarounds, but also eliminates tedious hum and noise troubleshooting and offers complete ground isolation.

Ethernet systems use standardized networking technology that provides transport of uncompressed and compressed digital audio through standard Ethernet hardware and media.

Copper cables are prone to problems because audio sources are very low voltage AC circuits and susceptible to ground loops caused by current flowing through the shield. Check soldered shields and inner conductors. Any break in the continuous flow of audio from connectors or patch cords presents the possibilities of signal flow or grounding problems. Patch bays are an often likely source of problems because of wear, corrosion and travel.

## Precabled Venues

It has become common for venues (facilities) that are heavily used by television broadcasters to be precabled or prewired. Typically the television truck will park in the loading dock and there will be an access panel with audio and video connection points. There are two methods used when a venue chooses a permanent cable installation. Some venues will locate a master panel, usually on the ground floor in the vicinity of the television compound, which has dedicated point-to-point wires from the master panel to various other panel locations in the venue.

The master panel will have all the inputs and outputs for all announce booths, fields of play, locker rooms, roof tops, hallways and usually about every place that a broadcaster may want to do a remote broadcast from. An NFL, NBA, or major-league baseball facility will have a minimum of three announce booths: home, visitors and radio, plus multiple sideline positions for announcers and field-of-play sound effects. In a football venue there will normally be access panels on both sides of the field and each end zone. A baseball stadium will have access panels to home plate, near each dugout, first and third base, bullpens, outfield and select locations to capture the atmosphere of the venue. The number of positions and panels may seem like a lot and it is usually sufficient for most regular-season play. However, if you factor in a Superbowl, World Series or NBA Championship with pregame, game, post-game and international feed, the number of available circuits are used in a hurry.

The most flexible method of wiring a venue is to have the various connection points flow through a central patching station. The Staples Center in Los Angeles uses a central patching room that interconnects all the various connection panels (media panel) throughout the entire complex plus serves as the interconnect for the inbound and outbound "Telco" feeds (Figures 6.4, 6.5, and 6.6). (Telco is an abbreviation for telecommunication, which includes telephone but more importantly the fiber-optic feeds.) The advantage of a central patching facility is maximum flexibility and usually reliability, but the downside is the additional personnel to connect and interface the various shows and broadcaster needs plus maintain the network. From an audio and broadcasters point of view, a maintained facility is the key to quality and reliability of the venue's connectivity.

**Figure 6.4** Greg Glaser at Staple Center patch central. Every panel in the Center can be patched to each other through this wiring cross matrix.

For many years, long distant audio circuits were rented from the telephone company. Pictured is a standard telephone "punch block" where a piece of wire is inserted (punched) into a locking connector of the punch block for connectivity. The punch block can accommodate up to 25 pairs of single-strand copper wire. This is a very fast, efficient and reliable way to handle a large bundle of wires.

In addition to a well-maintained facility, the patching center helps to insure an equitable distribution of connectivity to all the clients. In a hard-wired, nonpatching facility there are usually maintenance issues because an outside contractor probably did the installation and the venue usually does not have the proper staff to fix and maintain the panels.

**Figure 6.5** Staples Center remote panel. Each panel is wired to the central patch panel.

Unfortunately, various television technicians will be connecting and disconnecting into a central panel and the possibility of error or of someone disconnecting someone else is very high.

**Figures 6.6 (a)** Close-up of a patch panel in venue **(b)** 19-inch rack in venue breakout room.

# Fiber Optics

Corning and Bell Labs were some of the first to begin research into fiber optics in the 1960s and in those days it required a Ph.D to install it, while copper wire was easy to install. The first major installation was in Chicago in 1976 and, by the early 1980s, fiber networks connected the major cities on each coast and began replacing telco copper, microwave and satellite links. Fiber-optic audio snakes appeared with great fanfare in the late 1980s at an AES convention and have been gradually replacing passive copper snakes across the spectrum of applications.

Fiber-optics technology works by sending signals of light down hair-thin strands of glass or plastic fiber. Light pulses move easily down the fiber-optic line because of a principle known as "total internal reflection." The light transmitted over the optical fiber is reflected along it like a cylindrical mirror. This phenomenon works because the angle of reflection does not allow the light to get out of the glass. The input side "muxes" or encodes the signals. ("Mux" is an acronym for multiplex, which is the process that converts the analog signal to a digital stream.) The output master unit demuxes the signals and converts the channels back to analog.

The optical transmitter processes and encodes the electronic pulse into an equivalent optical series of light. The transmitter is composed of a laser diode and a driver. The role of the driver is to convert an incoming electrical signal into an equivalent light signal and a lens focuses the light waves into the fiber-optic medium, where they transmit down the line. The light source is pulsed on and off, and a light-sensitive optical receiver on the other end of the cable converts the pulses back into the digital ones and zeros of the original signal. More sophisticated laser drivers have control circuitry for automatic power and temperature control.

The optical receiver converts the incoming optical power signal into an output data signal. The optical receiver is composed of three main parts: a photodiode, an amplifier and decision circuitry. The photodiode will convert an optical signal into a current signal. The current signal is fed into an impedance amplifier, where it is converted and amplified into a voltage signal.

**Figure 6.7** ST type fiber connector.

There are three types of fiber-optics cable commonly used: single mode, multimode and plastic optical fiber (POF). Hair-thin fibers consists of two concentric core layers of high-purity silica glass, a cladding, and a protective sheath. Light rays are modulated into digital pulses and travel along the core without penetrating the cladding because the cladding bends the light. A mode is a defined path in which the light travels. A light signal can propagate through the core of the optical fiber on a single path (single-mode fiber) or on many paths (multimode fiber). The mode in which light travels depends on geometry, the index profile of the fiber, and the wavelength of the light.

Single-mode fiber has the advantage of high information-carrying capacity, low attenuation, and low fiber cost, while multimode fiber has the advantage of low connection and electronics cost, which may lead to lower system cost. Most fiber-optic systems can carry signals for distances of a mile or more on relatively inexpensive multimode fiber. Single-mode fiber can go for many miles but is much pricier. Generally speaking, the "pure" fiber systems have lower latency, although new proprietary audio networking protocols have reduced latency to the one mile range or below, making pure fiber systems effectively real-time in most live sound applications.

Single mode cable is a single strand of glass fiber with only one mode of transmission. Single mode has a relatively narrow diameter and a smaller core than multimode cable. With single mode cable, the light travels nearly parallel to the axis of the cable, creating little pulse dispersion and the least amount of signal attenuation. Single-mode fiber gives a higher transmission rate, higher bandwidth and up to 50 times more distance than multimode, but requires a light source with a narrow spectral width.

Multimode fiber gives more bandwidth at higher speeds over medium distances. The greater capacity is beneficial for many applications. However, light waves are dispersed into numerous paths, or modes, as they travel through the cable's core. In long cable runs (greater than 3000 feet (1000 meters) the multiple paths of light can cause signal distortion at the receiving end, resulting in an unclear and incomplete data transmission.

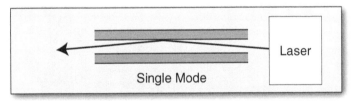

**Figure 6.8** Single mode fiber. The small core and single light-wave virtually eliminate any distortion that could result from overlapping light pulses, as in multimode fiber.

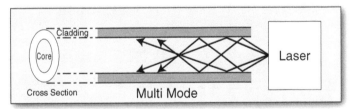

**Figure 6.9** Multimode fiber.

Even laser light shining through a fiber-optic cable is subject to loss of strength, primarily through dispersion and scattering of the light within the cable itself. The faster the laser fluctuates, the greater the risk of dispersion. Light strengtheners, called repeaters, may be necessary to refresh the signal in certain applications.

# Fiber vs. Copper

Copper cables can pick up interference from electromagnetic and radio frequencies (EMI/RFI), where fiber-optic cable benefits from total immunity to signal interference. Electromagnetic (EMI) and radio-frequency interference (RFI) are common in electrical power sources, long cable runs and lighting systems in live theatre and music productions. Fiber-optics are replacing weighty multicore cable with a thin strand of glass cable that can carry dozens of audio channels for many miles with absolute freedom from signal degradation, ground loops, or electrical interference. With long copper cable, there is more resistance and greater capacitance in the cable, which creates a low-pass filter lowering the cut-off frequency.

While fiber-optic cable itself has become cheaper over time, an equivalent length of copper cable costs less per foot but not in capacity. Fiber-optic cable connectors and the equipment needed to install them are still more expensive than their copper counterparts.

## Broadcast Systems

A typical broadcast installation uses a combination of copper wire and fiber optics. Copper cable like coax, microphone mults and single microphone cables are common for localized interconnect and interface. Integrating analog and digital systems and preventing the signal from degrading through excessive analog-to-digital-to-analog conversions is a challenge.

**Figure 6.10** Cable from venue waiting to connect to OB van.

Audio systems design is changing the signal flow and processing component of audio management. Mixing consoles are becoming control surfaces, which has resulted in a shift from using microphone preamplifiers in the high-priced mixing consoles to using stage or field boxes that have quality preamplifiers and A/D/D/A conversion. Digital data streams are easily routed to through a matrix hub to digital consoles and other audio-signal users. Centralized signal flow facilitates routing and processing.

**Figure 6.11** Fiber headend in compound at Daytona.

## Digital Mults

Copper or fiber cable is the transport or carrier of the electrical or light pulses, but it requires the encoders and decoders that couple the fiber or copper together to make a quality and usable system. Input and output interfaces using optical-fiber transmitters have been developed in response to a growing demand for a flexible method of transporting signals. Broadcasters require very high speeds for digital video and audio over long distances and wanted to reduce the weight and dependence on copper of remote telecasts. Broadcasters use an extensive range of converters, transceivers, multiplexers, and other associated components for combined audio/video/data systems down one fiber.

Initial fiber systems had problem with dynamic range in addition to having the gain control adjustments at the stage I/O or control box. Early fiber-optic snakes used 16- and 18-bit converters but needed more limiting to help in handling peaks. Broadcast systems use 24-bit converters and many systems offer sampling rates of up to 96 kHz or even 192 kHz, the equivalent of pristine-quality recording audio.

DSP technology allows software control of audio engine, status monitoring, channel control to reconfigure signal routing, mic preamp gain, limiter functions, phantom power, and sampling rate conversion. A/D conversion is 24-bit with sampling rates selectable from 44.1 kHz to 96 kHz. Digital mults place the microphone preamplifiers as close to the microphone as possible. Aphex Electronics makes microphone preamplifiers that are close to the source and can be controllable remotely from the mixer.

Broadcast systems include combinations of audio, video and data signals. There are analog and digital systems representing various price points and capacity. Audio modules accommodate analog mic/line inputs; SDI; AES/EBU digital I/O; intercom (RTS, Clear-Com, 4W); serial data; GPI paths; and Ethernet. Digital systems use a protocol language that transports packets of data between the I/O boxes. Note that proprietary audio-networking protocols are not compatible with each other. The Lightwinder-Natrix architecture and Klotz's Vadis software allow independent DSP control and remote assignment of any input to any output or any combination of channels.

Systems range from up to 256 AES digital channels or 512 analog audio channels on one fiber with latency specified of less than 500 microseconds.

**Figure 6.12** Interconnect from fiber breakout to copper mults. Remote patchbay interface.

Most systems provide dual transmission lines that connect the units for redundancy along with coaxial inputs and outputs. Additionally, each I/O system frame has redundant power supplies and RS-422 interface for communication and machine control; external word clocks at all sampling frequencies; and I/O for master or slave mode.

Ethernet-based digital audio systems using standard CAT-5 cabling are frequently used for live sound and have possibilities for broadcasters. Copper-based Ethernet systems are used to transport digital audio signals on CAT-5 connections when cable runs do not exceed approximately 328 ft maximum range without repeaters. There are a variety of audio-specific encoders and protocols providing a range of capacity over existing bandwidth and minimum signal latency. CobraNet was the first audio protocol followed by long-time cable manufacturer Whirlwind offering analog or digital inputs and outputs with 32×32 channels. As with most digital systems, switching and routing is done using software. Ethernet-based systems will advance with improvements in audio encoding and transport architecture that will increase channel capacity and usable lengths.

## Analog Audio

*The advantage of analog audio is that the signal can be transported without being processed. The single variable in preventing signal degradation is a solid connection between equipment so that the audio signal can transfer from one conductor to another.*

**Figure 6.13** Punch blocks in compound at Daytona.

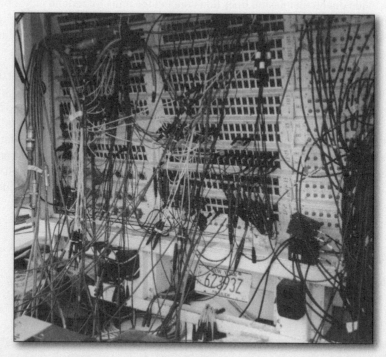

**Figure 6.14** CBS OB van used panel-mounted XLR connectors in the audio access panel.

## Signal Delay or Latency

Analog-to-digital-to-analog conversion introduces a delay into the audio and video signals. This delay or latency will cause synchronization issues with video, which must corrected. A/D-D/A conversion can add up to 2 ms depending on sampling rate and each additional DSP device in the audio chain adds its own A/D-D/A conversion and latency unless the signal stays in the digital domain.

Latency also stems from the conversion of data into packets for network transport. Packets of data contain a certain number of audio samples and the system must store ("buffer") the samples while it builds packets. Optimum transfer is a tradeoff between transmitting several small packets or fewer large ones.

Cable length is an insignificant factor unless distances are on the order of kilometers or miles. And even then, it takes one mile of copper or fiber to introduce even 0.02 ms of latency. Cable latency can be caused by propagation delay, switch delay from buffering and forwarding delay caused by network switches and by the size of the data pack and bandwidth of the network.

Ethernet switches, for example, can add 5 to 120 μs, depending on frame length and whether it's 100 Mb or 1G Ethernet. CobraNet has selectable latency modes: 5.33 ms, 2.66 ms or 1.33 ms. For Ether-Sound, latency is a constant 1.25 μs (6 samples @ 48 kHz) plus less than 2 μs per additional node in a daisy chain, not dependent on the number of channels.

# Handling Cables

Cables are easily damaged from lawn mowers and spikes and are vulnerable on the golf course or field. Most fiber manufacturers provide a plastic boot that fits over the ferrule body for protection.

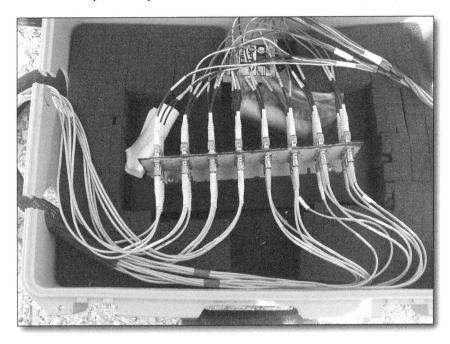

**Figure 6.15** Extending the fiber cables in a protected junction box.

When using interconnection housing to mate two optical fibers, it is good practice to remove dust particles from the housing assembly with a blast of dry air. Larger dust particles can totally obscure light. Whenever a fiber is unmated it must be covered immediately. Every time that a fiber is mated or unmated it must be cleaned. This is important, since dust particles on the ends of the optical fiber can add up to 1 dB of loss.

## Handling and Connecting the Fibers

Transmission characteristics of fiber are dependent on the shape of the optical core and are affected by abrupt bending. A suggested minimum bending radius is 3 cm—smaller than this can lead to bending loss, decreasing the available power. Additionally, do not place fiber-optic cable under heavy objects.

## Fiber Connector—Preloaded Epoxy or No-Epoxy and Polish–ST, SC, and FC Connector Styles

Installation of connectors requires training and skills, especially for polishing and curing the epoxy. There are connectors that use epoxy, that do not use epoxy, and that have preloaded epoxy. There is an initial set-up time for the field technician, who must prepare a workstation with polishing equipment and an epoxy-burning oven. The termination time for one connector is about 25 minutes due to the time needed to heat cure the epoxy. Average time per connector in a large batch can be as low as 5 to 6 minutes. Faster curing epoxies such as anaerobic epoxy can reduce the installation time, but fast-cure epoxies are not suitable for all connectors.

Connectors that are preloaded contain a measured amount of epoxy. These connectors reduce the skill level needed to install a connector, but they don't significantly reduce the time or equipment needed. The preloaded epoxy connectors require the same amount of installation time as standard connectors, usually 25 minutes for the first connector and 5–6 minutes average for a batch.

Some connectors do not use epoxy because they use an internal crimp mechanism to stabilize the fiber. Connectors that use the internal crimp method install in 2 minutes or less. The crimp connector is more expensive to purchase than a standard connector, but requires less training since the skill demands are reduced (learning the crimp mechanism is easier to master than using epoxy). Consumable costs are reduced to polish film and cleaning supplies.

The straight tip (ST) connector is a popular fiber-optic connector originally developed by AT&T. The subscription channel (SC) connector is a push-pull type of optical connector that features high packing density, low loss, low back-reflection, and low cost. (See Figure 6.16.)

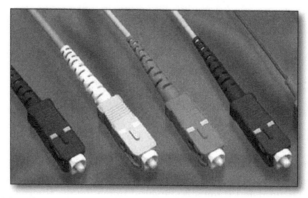

**Figure 6.16** SC type connectors.

## *Cleaning Optical Fibers*

Optical fibers must always be cleaned before mating. The required equipment is Kimwipes or any lens-grade, lint-free tissue, denatured alcohol and canned dry air.

The technique is to fold the tissue twice so it is four layers thick and saturate the tissue with alcohol. First clean the sides of the connector ferrule. Place the connector ferrule in the tissue and apply pressure to the sides of the ferrule. Rotate the ferrule several times to remove all contamination from the ferrule sides. Now move to the clean part of the tissue. Be sure it is still saturated with alcohol and that it is still four layers thick. Put the tissue against the end of the connector ferrule. Place your finger against the tissue so that it is directly over the ferrule. Now rotate the end of the connector. Mate the connector immediately. Do not let the connector lie around before mating. Dry air can be used to remove dust from the connector housing or the transmitter or receiver ports.

## *Safety*

Never look directly into the optical fiber. Nonreversible damage to the eye can occur in a matter of milliseconds. Never touch the end face of an optical fiber.

# Digital Equipment

Equipment manufacturers produce a range of specialized converters, transceivers and multiplexers for audio systems or video-based systems. Audio-based systems transport audio, intercom and data signals, and other associated components for combined audio/video/data systems down one fiber. It is possible to combine multiple signals onto a single wavelength of light. For example, two video signals traveling in opposite directions can share the same wavelength as long as the lasers used have directional isolation.

Video systems transport HDTV and SDTV video signals with embedded AES audio channels, remote control, Ethernet computer networking and data onto a single fiber. Video requires a considerable amount of bandwidth and a single fiber will carry far fewer video signals than audio. Video-based systems use audio embedders and de-embedders that combine audio and video signals, which are modulated into a single wavelength in the electrical-to-optical converter. Embedders are also used for data that is often needed to be unidirectional.

Klotz Digital was one of four companies that introduced digital audio snakes in 1969 with the company's Oaklink system. The system is long gone, but Klotz remains active in DSP processing and digital audio networking with its broadcast-targeted Vadis digital audio and control network.

The Telecast Adder System is proprietary architecture, as are most with input modules for analog mic/line signals; AES/EBU digital inputs and outputs; intercom (RTS, Clear-Com, 4W); serial data; GPI paths; and Ethernet. It has up to 256 AES digital channels or 512 audio channels on one fiber, with 500 microseconds of latency.

Otari manufactures the Lightwinder system, which has been used in many live sound applications where long runs and interference make quality audio a challenge, including the 2000 Sydney Olympics, NFL's Superbowl, World Cup Skiing, concerts and live theater productions. Otari Lightwinder can provide up to 64-channel, bidirectional capacity. A/D conversion is 24 bit with sampling rates selectable from 44.1 kHz to 96 kHz.

The Reidel company of Germany designed a stage box system for the Yamaha digital mixing console connecting over distances of up to 1,640 ft with multimode fiber providing up to 80 channels of bidirectional audio and a SCSI bus for insert points. Reidel has made a name for itself in communication systems but essentially has developed a large-scale audio matrix with high-fidelity audio inputs and outputs. This matrix can be configured to act as an audio router, communications network or Reidel can operate on a fiber and/or CAT-5 network.

Systems can be custom-assembled in a modular approach from separate component cards and installed in a rack-mount card tray. Many functions are available and engineered as combined audio-video systems, and audio and video cards can be mixed in the same tray. These can be used as stand-alone units for specific applications.

System designs generally optimize bandwidth to perform a range of specific and related functions. For example, audio systems provide microphone amplifiers, phantom power and phase reversals for each channel input, which would not be included in a video or line-level audio transport system. Video requires significant bandwidth and provides between two and eight channels of audio to maintain synchronization, but generally does not include DSP or other audio functions.

## Integrating Analog and Digital Systems

Interconnectivity b etween analog and digital equipment requires analog-to-digital-to-analog conversions. Fiber optics only transports digitized signals, which basically presents a problem.

**Figure 6.17** Broadcasters who have complex remote announce booths use fiber optics and remotely powers PL's and IFB's.

The FOX NASCAR announce booth box is a package of IFB control strips, microphone preamplifiers, audio PL strip, fiber en/decoder, two-wire to four-wire converters, power supply for stage manager, spotters, and statistician. The box will also have a UPS power back-up to keep it operational in the event of a power failure.

IFBs and PLs are remotely powered and are far less susceptible to interference and distance-related problems. PLs need four-wire converters because the communication system passes its signal through the fiber optics as a CAT or through its own fiber. At that point, the communications is four-wire with a dedicated talk and listen wire. These signals need to be converted to a powered bidirectional communication circuit through the two- to four-wire converters and then to a power supply for beltpacks.

Microphone signals are sent to the OBV line level. Microphone preamplifiers are as close to the microphone as possible. Aphex Electronics makes a microphone preamplifier that resides close to the source and is controllable remotely from the mixer.

Fiber systems for entertainment and broadcast are increasing using remote-controlled preamps to the announce and stage area. This makes a line-level fiber-optic snake more attractive in high-quality systems where audio headroom and bandwidth are required. Component reliability has been proven in countless applications and generally redundancy is built into the design.

# **7** Microphones

The task of sound for television and film is to enhance the picture and enrich the listening experience of the viewer. Film sound is a process of accumulating and building soundtracks, then mixing and building a soundscape that appropriately matches the picture. Sound for television is a similar building process, but the execution and delivery is in real-time through a mixing desk with properly placed microphones to pick up the desired sounds. In live television, all sound elements are captured and combined as they occur.

Microphones are the fundamental tools used in sound design, recording and reproduction and are essential equipment in capturing dialog, effects and even the sound of the drums in the band playing the bumper music as the show goes to break. In the sports world microphones are an often misunderstood piece of equipment, because good microphones are expensive and coveted. When the freelance television trucks started rolling in the late 70s, they carried six shotgun microphones, one for each camera, four laveliers, four hand microphones and four announcer headsets—and that was it!

Freelance audio mixers usually had to work with what microphones were available on the television truck, which was often inadequate for a lot of television productions. It became common for some of the top sound mixers to carry microphones with them or rent specific ones for a production.

**Figure 7.1** Shotgun microphone properly mounted on large-lens camera. The mounting clamp was custom made by Audio Technica for the 2000 Summer Games in Sydney, Australia.

In the early days of broadcasting, microphones were generally large in size and adopted from film and radio. Shotgun microphones on large mechanically moveable arms called a "perambulator" permitted movement on a large stage and kept the microphone out of the camera eye while still capturing strong on-presence dialog (Figure 7.1). Live television in front of a studio audience brought plenty of challenges but as television went mobile, audio had to pioneer new frontiers.

Audio finally became recognized as a critical production element and networks like FOX, NBC and ESPN began hiring audio consultants to improve the quality of their audio. Bob Dixon is responsible for every aspect of the sound for NBC Olympics and Ron Scalise oversees sound for all ESPN Networks and they are concerned with the sound from capture to consumers. Fred Aldous and Denis Rhyan elevated the sound of NASCAR to new standards and are focused and committed to quality sound, which includes using a wide variety of high-quality microphones.

Surround sound has put new demands on sound production, and more consideration must be given to microphone selection and placement.

**Figure 7.2** A hypercardioid microphone (AT3000) mounted very close to a snare drum reduces the sound from the surrounding drums and cymbals.

Entertainment productions like awards shows generally have adequate budgets and top mixers like Ed Greene and Klaus Landsberg insure high standards for sound quality. At the 2005 Grammy awards, over 40 microphones were used for the atmosphere mix.

Microphone design has not only benefited from technology advances and cheaper production costs, but also from fierce competition from microphone manufacturers, who have in the last ten years significantly expanded the range and quality of microphones and transducers that are available and used by sound engineers. The old notion that a microphone belongs in the studio or in a particular application has disappeared and been replaced with a strong desire for sonic creativity that motivates the sound engineers to push the boundaries of conventional noise.

Microphone placement was the golden rule when I began my audio career, along with many other studio engineers, by moving microphones around for the chief engineer. I remember reading articles about engineers spending hours on microphone placement for a drum kit. This may seem ridiculous. However, if you examine the electronic characteristics of microphones and how placement of the microphones interacts with acoustics, then you can understand why it may have taken so long. Proper microphone selection and placement is critical because, once competition or rehearsals begin, it is often impossible to change microphone positions. If you make a poor selection or bad placement, the results can be very unforgiving.

The general thought in television sound design is to use an extensive combination of microphones on cameras, up close to the action and at various locations to pick up the atmosphere of an event. In surround sound, a lot of attention is paid to capturing the atmosphere of an event to give some spatial orientation to the picture. Close microphone placement and microphones on handheld cameras make audio very "point of view" (POV) oriented and requires an aggressive mix for audio to match the visuals. An aggressive mix is subjective as well, but when too many microphones are open in the mix, or when the wrong microphones are in the mix, the clarity and definition of the mix are buried or lost. To achieve a quality sound track with good sonic qualities, some microphone planning is essential, because it is the microphone that sets the quality level of the sound reproduction.

Here are some pointers on microphone selection and placement:

1) Examine the acoustic content of the sound source. The sound pressure level, dynamic range and frequency characteristics of the sound source are the guides to the microphone selection.

2) Examine the physical characteristics of the sound source: Is the source moving, how close can microphones be placed to the sound source, or can a microphone be attached to the source?

3) Examine the acoustic characteristics of the environment. Is there excessive ambient noise, or is it highly reverberant?

4) Most importantly, listen! Like a musician, sound engineers must train their ears and know how to listen and what to listen to.

Television miking has been handed down, shared and exchanged amongst a handful of audio mixers. Considering the first Super Bowl was televised in 1967 by both CBS and NBC, television coverage and sound has come a long way.

## The Basics of Microphones

The microphone is the first piece of equipment in the audio signal chain and has the job of converting acoustical sound waves into electrical images of the sound. The correct choice and placement of microphones is essential to establish a proper foundation for the sound mixer to build the soundscape.

A microphone is known as a transducer because it converts acoustical energy to electrical energy. A microphone operates by using a diaphragm to capture the motion of air pressure variations caused by sound waves. The sound wave causes a back-and-forth motion of the diaphragm, which is designed to harness and convert this mechanical motion from the change in air pressure on the diaphragm to an electrical image of the sound wave.

The direct sound strikes the front surface of the diaphragm, but if the diaphragm is exposed on both sides, then sound waves will affect the backside of the diaphragm. The way the diaphragm is exposed and suspended, along with the microphone housing, determines how the microphone responds and captures the sound wave.

A basic pressure-sensitive microphone design has only one side of the diaphragm exposed and is mounted with the back enclosed so the sound waves can approach the diaphragm from only the front direction (Figure 7.3). This approach is inherently omnidirectional but it gives a very even response to all frequencies.

**Figure 7.3** Omnidirectional capsule with only one side of the diaphragm exposed.

Another design allows the sound to approach the capsule from all directions. This gives microphone designers access to the area behind the diaphragm, where they use sound cavities to introduce front-to-back delay and acoustic cancellation to give the microphone a "directional" effect. How the microphone capsule responds to sound coming at it from all directions defines the directional characteristics of a microphone. Microphones are identified by these directional characteristics, which is how much sound they pick up or reject from various directions.

## Omnidirectional Microphones

While they must be used close to the sound source, omnidirectional microphones have a reduced sensitivity to noise and breath blasts, making them ideal for many clip-on mic applications. If the distances are equal, omnidirectional microphones work almost as well when pointed away from the subject as pointed toward it. This is why you often see lapel microphones pointed away from the mouth when breath blasts are excessive (Figure 7.4).

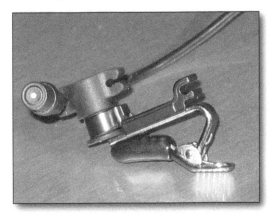

**Figure 7.4** Omnidirectional lapel microphone on tie clasp.

Omnidirectional microphones are normally better at resisting wind noise and mechanical or handling noise than directional microphones. Omnis are also less susceptible to "popping," caused by certain explosive consonants in speech, such as "p," "b"" and "t."

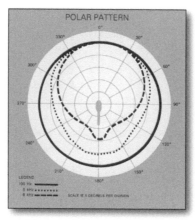

**Figure 7.5** Omni polar pickup pattern. The solid line shows equal sensitivity in all directions.

Omnidirectional microphones pick up sound equally from just about every direction (Figure 7.5). This equal sensitivity delivers a fairly uniform sound color and is said to be "on-axis." However, even the best omni models tend to become directional at higher frequencies, so sound arriving from the back may seem a bit "duller" than sound from the front, although apparently equally loud.

On-axis sounds differ in tone from off-axis sounds, because acoustical or electrical cancellation reduces the apparent level of certain frequencies. The sound is still audible but not full frequency. The shotgun microphone has the greatest off-axis rejection of sound and must be aimed at the desired sound. (See Figure 7.6.)

**Figure 7.6** Shotgun microphone on a boom pole. The shotgun help reduce reflective sound in a small dressing room with parallel walls.

### Microphone Response Pattern

*Microphone specifications are a technical description of how a microphone performs under certain conditions. Most tests are conducted in anechoic chambers, which is not the operational environment of the microphone.*

*A polar pattern is used to represent and to help visualize a microphone's directional sensitivity. These round plots show the relative sensitivity of the microphone (in dB) as it rotates in front of a fixed sound source. Printed plots of the microphone polar response are usually shown at various frequencies because a microphone does not respond uniformly to all frequencies.*

*The polar pattern represents a horizontal "slice" through the pickup patterns, while Figure 7.5 shows a full 360-degree representation of the directional characteristics of a microphone. Polar patterns are measured in an anechoic chamber, which simulates an ideal acoustic environment with no walls, ceiling, floor or reflected sound. In the real world, walls and other surfaces will reflect sound quite readily, so that off-axis sound can bounce off a nearby surface and right into the front of the microphone. See Figure 7.7 for representative polar pattern.*

The physical size of the omnidirectional microphone has a direct bearing on how well the microphone maintains its omnidirectional characteristics at very high frequencies. The body of the microphone simply blocks the shorter high-frequency wavelengths that arrive from the rear. The smaller the diameter of the microphone body, the closer the microphone can come to being truly omnidirectional at all frequencies.

## Directional Microphones

The pickup pattern of most omnidirectional microphones is roughly a circular pattern, while directional microphones exhibit a heart-shaped polar pattern and are called *cardioid* microphones (see Figure 7.7). Directional microphones are seen in a number of variations, such as cardioid, supercardioid, subcardioid, hypercardioid and bidirectional. They differ in their rejection of sound coming in from the sides and behind, but also in how they affect the pureness of the sound around the microphone. The goal of directional microphones is to increase sensitivity and reduce the reflective field. This effect also varies with frequency; only higher quality microphones are able to provide uniform rejection over a wide range of frequencies.

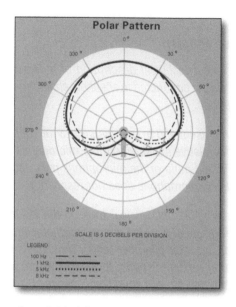

**Figure 7.7** Polar representation of cardioid pickup pattern.

This directional ability is usually the result of external openings and internal passages in the microphone housing that allow sound to reach both sides of the diaphragm in a carefully controlled way. Sound arriving from the front of the microphone will aid diaphragm motion, while sound arriving from the side or rear will cancel diaphragm motion.

**Figure 7.8** Omnidirectional microphone (left) and unidirectional (cardioid) microphone (right).

Improving a microphone's directional characteristics will improve the clarity of the audio. A typical cardioid microphone has a directional response, coverage or acceptance angle of approximately 120°, which is the hemispherical area reaching forward. Sounds from the rear are generally attenuated by some 30 dB, but this attenuation is dependent on the frequency. Even though the sound is attenuated to the sides and the rear, that sound will still affect the overall sound and could make the reproduction muddy.

In a podium application, directional or cardioid types of microphones are a good choice. By pointing the microphone directly at the desired sound source and having the unwanted noise positioned in the null (minimum point) of the pattern, the result is improved intelligibility of speech at a greater working distance, reducing feedback and echo. Cardioid microphones can help reduce unwanted sound, but rarely can they eliminate it entirely. When the null of the microphone is facing any unwanted sound such as other instruments or feedback from a sound reinforcement loudspeaker, the problems will be reduced.

## *Hypercardioid*

The hypercardioid microphone is more directional than standard cardioids but less directional than shotguns. Hypercardioids are not shotguns. They have more side rejection than cardioids but not as much as shotguns.

The distance factor for a hypercardioid is approximately double that of an omnidirectional microphone. This means that if an omni is used in an identical environment to pick up a desired sound, a cardioid can be used approximately twice as far from the sound source. The pickup pattern for a hypercardioid is shown in Figure 7.9.

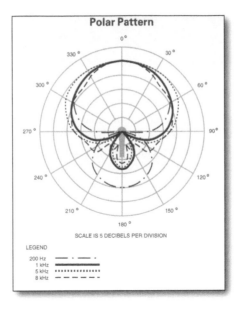

**Figure 7.9** Hypercardioid pickup pattern.

When hypercardioids are placed properly, they can provide more focused pickup and less room ambience than the cardioid pattern, but they have less rejection at the rear: −12 dB for the supercardioid and only −6 dB for the hypercardioid. Hypercardioids have a little "tail" of stronger sensitivity directly behind the microphone This can cause problems if that tail points at a noise source. Usually the dead zone on these is roughly 30 to 45 degrees off the straight-back centerline.

**Figure 7.10** A dual capsule hypercardioid microphone design.

The distance a microphone has to be placed from the sound source is often determined by television's desires not to see a microphone in the camera shot or by safety issues with an object (microphones) on the field of play. A directional microphone can work at greater distances and may be the only choice when tracking a moving object.

## Shotgun

When miking must be done from even greater distances, line+gradient or *shotgun* microphones are often the best choice. Line microphones are excellent for use in video and film, in order to pick up sound when the microphone must be located outside the frame—that is, out of the viewing angle of the camera.

By exposing both sides of the diaphragm, the capsule is reacting to the difference in pressure of both sides. This pressure gradient approach has definite directional characteristics and various designs utilizing electrical and acoustical summing have been used. Shotgun microphones combine a directional or gradient element with an interference tube that utilizes acoustical vents and screens to increase cancellation at the rear. The interference tube is in front of the element to ensure much greater cancellation of sound arriving from the sides.

The tube causes any sound energy arriving from the sides to undergo partial cancellation before it can reach the capsule. The exact degree of this cancellation depends greatly on the wavelength of the sound. For wavelengths at low and midrange frequencies, longer than the tube, the interference tube has little effect. See Figure 7.11 for some examples.

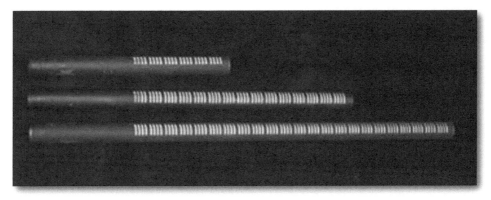

**Figure 7.11** Shotgun microphones come in a variety of lengths used for the interference tube.

As a general design rule, the interference tube of a line microphone must be lengthened to narrow the acceptance angle and increase the working distance. While shorter line microphones may not provide as great a working distance as their longer counterparts, their wider acceptance angle is preferred for some applications, because aiming does not need to be precise. Audio Technica's professional shotgun microphones employ an exclusive design (U.S. Patent No. 4,789,044) that provides the same performance with an interference tube one-third shorter than conventional designs.

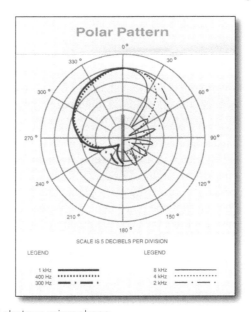

**Figure 7.12** Polar pattern of shotgun microphone.

At higher frequencies the pickup pattern becomes narrower but with great variations in response for different angles and frequencies, resulting in a more complex directional characteristic than ordinary polar diagrams. See Figure 7.12 for a polar pattern of a shotgun microphone. Figure 7.13 shows a stereo shotgun on a handheld camera.

**Figure 7.13** Stereo shotgun microphone on handheld camera.

A shotgun microphone must be well aimed at the intended sound source, and that source must fit within the microphone's front pickup angle. Otherwise, comb-filter-like effects will result from any reflected sound, off-axis sound sources, or motion of the direct sound source (or of the microphone itself, if it is used in a reverberant space). The best situation to use a shotgun microphone is outdoors where there is little reflected sound energy and limited off-axis sound.

### Distance Factor

A directional microphone's ability to reject sound that arrives from off-axis provides a greater working distance or "distance factor" than an omni. A cardioid microphone may pick up satisfactorily at 3 feet and an in-line shotgun microphone may reach between 6 to 9 feet.

Pickup distance is very difficult to define and will be dependent on your signal-to-noise ratio. The sound source or signal will have a loudness level that, at a certain distance from the microphone, will be the dominant sound and mask any interfering sound. For example, a race car has a sound pressure level of over 120 decibels. It is not necessary to place the microphone one foot away from the automobile to capture the sound. A more distant microphone placement allows the microphone to pick up additional reflective sound waves which also captures the acoustic dynamics of the event.

## Which Pattern Is Best?

Whether you should select a directional or omnidirectional microphone will depend on the application, the acoustic environment, the working distance required and the kind of sound you wish to capture. Directional microphones can suppress unwanted noise, reduce the effects of reverberation and increase gain-before-feedback. But in good acoustic surroundings, omnidirectional microphones, properly placed, can preserve the "sound" of the recording location, and are often preferred for their flatness of response and freedom from proximity effect.

Directional microphones may need to compensate for the bass loss and other off-axis coloration. Additionally, some microphones have noticeably poor off-axis response, which means that sounds entering the microphone from the sides and the rear are strongly colored.

From a distance of two feet or so, in an absolutely dead room, a good omni and a good cardioid may sound very similar. But put the pair side-by-side in a live room, such as a large church or auditorium, and you'll hear an immediate difference. The omni will pick up more of the reverberation and echoes—the sound will be very "live." The cardioid will also pick up some reverberation but a great deal less, so its sound will not change as much compared to the dead room sound. If you are in a very noisy environment, and can point the microphone away from the noise, a comparison will show a better ratio of wanted-to-unwanted sound with the cardioid than with the omni.

## Small-diaphragm Microphones

Small-diaphragm microphones have been around for awhile and some close miking techniques originated with the *lapel* microphone, which is found on all news and talk show sets. An innovative sound engineer moved those lapel microphones off the announcers and onto the field of play, where small microphones are placed in the nets of tennis, volleyball, table tennis and badminton as well as fixed to the basketball backboard and placed inside baseball bases. Additionally, lapel microphones have been placed on all bars at gymnastics, giving a very close perspective of the athlete against a noisy, large, enclosed hall (Figure 7.14).

The practice of using small microphones as close to the action as possible has gained acceptance because the newer smaller microphones have high-quality sonic characteristics and minimum off-axis coloration. Close miking delivers a very high signal-to-background noise and reduces reflective and reverberant sound, which can make a mix muddy.

**Figure 7.14** Rigging miniature microphone on gymnastics bars.

Omnidirectional condenser microphones have good extended low-frequency response and lower distortion than directional microphones in a distance of over 30 cm. Omnidirectional microphones can be moved closer to the sound source without the proximity effect that occurs with a directional microphone. All omnidirectional microphones become increasingly directional for higher frequencies.

Small-diaphragm microphones usually use a stiffer diaphragm than a large diaphragm microphone and are able to handle relatively higher SPLs. Omnidirectional microphones offer high headroom before clipping and usually very low distortion specifications, which is especially important when working with high SPLs in close miking situations.

The lapel microphones have gotten smaller and the microphone capsule and diaphragm are always separate from any electronics or power supply. This minimizes the profile and acoustic influence of the housing and makes this a very useful microphone for extremely close miking.

## Safe Microphone Placement

*Microphone placement is governed by practicality and effectiveness in capturing the desired sound, but the utmost concern is for the safety for anyone around the microphone. It is not practical to put a microphone everywhere you would desire to have one. Shotgun microphones hanging from the ceiling are an effective way to capture the sound of boxing or ice hockey, because the microphones are pointing directly at the desired sound (on-axis sound) and the excessive crowd (off-axis sound) is not direct or as strong.*

*Safety is always a concern because microphones are rigid and often made of metal and could seriously hurt someone if fallen upon. Microphone cables should always be taped down where people are walking in order to avoid tripping. (I have heard a few stories about some embarrassing situations, but have never heard of any serious injuries to humans or animals.)*

*All organized sports have sanctioning bodies and federations that govern the rules of competition, event venues and sometimes even the participants. During the preparation for the 2004 Summer Games, I sent diagrams of all microphone locations to the Gymnastics Federation for approval. I had microphones within inches of all joints and connection points on the parallel and uneven bars and had used this approach successfully in a previous Olympics. The International Gymnastics Federation approved the entire plan without changes. However, just hours before the beginning of competition a Gymnastics Federation delegate decided that the microphone positions were not safe! I begged and pleaded, but I had to move all the miniature microphones sometimes twice as far as was originally agreed to. (See Figure 7.15.)*

**Figure 7.15** Miniature cardioid microphone on gymnastics.

## Large-diaphragm Microphones

The large diaphragm is easier to move, even with low sound pressure levels, and will therefore provide a larger output. The sensitivity and surface area of the larger, more compliant, diaphragm is generally higher than the small and stiffer diaphragm.

**Figure 7.16** AT4050 large-diaphragm microphone used in 2006 Olympic Games for surround channels.

Typical broadcast applications for this type of microphones have generally been for voiceovers, but an expanded view of this microphone found additional uses. Beginning in 2000 the Olympics began using these microphones for rear-channel surround and to capture the low-end frequencies of the pyrotechnics during the ceremonies (Figure 7.16).

# Microphone Characteristics and Properties

Most microphones are analog devices and, due to the nature of materials and design, all microphones will have certain acoustic characteristics that may color the reproduction of sound. All pressure microphones exhibit a phenomenon called *presence boost* where a certain presence peak occurs at high frequencies. This phenomenon occurs when the wavelength of the sound is comparable or smaller than the diameter of the diaphragm. The sound pressure waves are reflected off the surface of the diaphragm, creating a sound pressure build-up between incoming and outgoing sound in front of the microphone diaphragm. This effect causes an on-axis rise in output as the frequency goes up to a maximum of about 9 dB at 20 kHz for a half-inch diaphragm.

When you move a cardioid microphone closer to the sound source, the lower frequencies are boosted. This is known as the *proximity effect*. Too much proximity effect will cause a boomy sound and affect the tonal balance between low frequencies and the mid- and high-frequency range. Microphones can be acoustically corrected for a flatter response but sometimes an inconsistency like presence boost or proximity effect becomes a microphone's acoustic fingerprint, character and legend and is desirable. For example, for many sound engineers a certain microphone has a natural enhancement to a sound and the engineer has a preference for that effect.

The microphone capsule is assembled in package that makes it structurally useful and acoustically functional. The size of the microphone capsule and how it is positioned in the housing package has an effect on the sound of a microphone. An acoustic shadowing phenomena can occur around the microphone due to the shape and the size of the microphone body containing the preamplifier, the connector and the design of the protecting grill.

## Electrical Properties of Microphones

Microphones produce an electrical image of a sound wave and are classified by how they convert sound energy to an electrically equivalent signal. There are two predominant methods, electrodynamic and electrostatic, better known as *dynamic* and *condenser*.

Electrodynamic or dynamic microphones operate when sound waves cause movement of a thin mylar diaphragm. Attached to the diaphragm is a coil of wire that freely moves through a magnetic field. The mechanical movement of the diaphragm and the attached coil of wire through a magnetic field creates an electrical impression or electrical equivalent of the sound wave from the movement of air. The amount of electrical current is determined by the speed of that motion. The fundamental elements of a dynamic mic are shown in Figure 7.17.

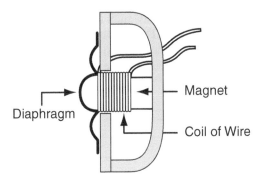

**Figure 7.17** Basic elements of a dynamic microphone.

The advantages of the dynamic microphone are its extreme reliability and ruggedness plus it is inexpensive for the sound quality obtained. They do not need batteries or external power supplies and their output level is high enough to work directly into most mixing console inputs with an excellent signal-to-noise ratio. They are capable of smooth, extended response, or are available with "tailored" response for special applications. They need little or no regular maintenance, and with reasonable care will maintain their performance for many years. The downside is the potential quality of the dynamic microphone is limited by how efficiently the sound moves the mass of the diaphragm.

**Figure 7.18** The hand microphone is used for the stand-up position at all sporting events. It can be a dynamic or condenser and is often an omnidirectional microphone.

Condenser (or capacitor) microphones use a lightweight flexible diaphragm, usually a very thin mylar film, coated on one side with gold or nickel. The diaphragm is mounted very close to a conductive stationary back plate that acts as the opposite side of a capacitor. Sound pressure against this thin polymer film causes it to move and changes the capacitance of the circuit, creating a changing electrical output. (See Figure 7.19.)

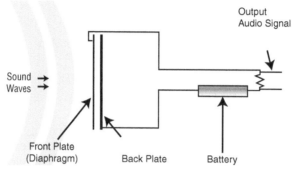

Sound Waves →

Output Audio Signal

Front Plate (Diaphragm)    Back Plate    Battery

**Figure 7.19** Basic condenser microphone.

Condenser microphones are preferred for their very uniform frequency response and ability to respond with clarity to transient sounds. Condenser microphones can be made much smaller (and less conspicuous) than dynamics without compromising performance. Because of their low-mass diaphragms, condenser microphones are inherently lower in handling or mechanical noise than dynamic microphones. The low mass of the membrane diaphragm permits extended high-frequency response, while the nature of the design also ensures outstanding low-frequency pickup. The resulting sound is natural, clean and clear, with excellent transparency and detail.

Condenser microphones either use an external power supply to provide the polarizing voltage needed for the capacitive circuit or the microphone has permanent charge applied when it is manufactured. The electret condenser microphone has its polarizing voltage impressed on either the diaphragm or the back plate during manufacture and this charge remains for the life of the microphone. The diaphragm and back plate are separated by a small volume of air to form the capacitor (or condenser).

**Figure 7.20** AT microphone on a landing mat at track and field.

When the diaphragm vibrates in response to a sound, it moves closer to and farther away from the back plate. As it does so, the electrical charge that it induces in the back plate changes proportionally. The fluctuating voltage on the back plate is therefore an electrical representation of the diaphragm motion. If you push the plate closer to the other, the voltage across the condenser will be larger. If the moving plate is further away, the voltage across the condenser becomes smaller. Because the diaphragm of the condenser is not loaded down with the mass of a coil, it can respond very quickly to transients.

Audio-Technica has elected to apply the polarizing voltage, or fixed charge, to the back plate rather than the diaphragm. By doing this, a thinner material can be used for the diaphragm, providing a considerable performance advantage over electret microphones of conventional design. Many Audio-Technica microphone diaphragms, for example, are only 2 microns thick (less than 1/10,000th of an inch)!

Condenser elements are an ideal choice for design applications because they weigh much less than dynamic elements and they can be much smaller. These characteristics make them the logical choice for shotgun microphones, lavalieres and miniature microphones of all types. A condenser microphone generally has better sonic characteristics, higher sensitivity for better pickup at greater distances, lower handling noise than dynamics, and extended frequency response. Condenser microphones have superior transient response for accurately reproducing sudden sonic impulses such as those produced by voice, piano and percussion.

**Figure 7.21** Miniature hypercardioid microphone on track.

Externally polarized or *discrete* condenser microphones seldom have internal battery power. Instead, a phantom power source is used to provide both the polarizing voltage for the element and to power the impedance converter. The electret condenser microphone doesn't need a power supply to provide polarizing voltage but an polarizing voltage but an impedance-matching circuit inside the microphone does require some power. This may be supplied by a small low-voltage internal battery or by an external phantom supply.

Phantom powering supplies a DC voltage to the microphone through the same shielded two-conductor cable that carries the audio from the mic. The phantom power may be supplied either by the mic mixer or from an external supply that is inserted into the line between the microphone and mixer input. For phantom power to function, the line between the power supply and the microphone must be balanced to ground, and uninterrupted by such devices as filters or transformers, which might pass the audio signal but block DC. Phantom power also requires a continuous ground connection (Pin 1 in the XLR-type connector) from the power supply to the microphone. Balanced-output dynamic microphones are not affected by the presence of phantom power, since there is no connection between the shield and either signal lead and, therefore, no circuit for the DC voltage.

A condenser microphone's capability of handling large sound pressure levels (SPLs) is limited by the distance between the diaphragm and the back plate and the rigidity of the diaphragm. This sets limits for how much a diaphragm can move before the distortion is too high. The power supply for the microphone preamplifier sets limits for the amount of signal that can be handled before clipping occurs.

To achieve extremely high SPLs, some manufacturers such as Danish manufacturer DPA use higher voltages on the pre-polarized back plate charged to approximately 230V, which allows us to move the diaphragm and the back plate further away from each other without losing sensitivity. Hereby the diaphragm is able to have a bigger displacement without touching the back plate, which would make the signal clip. The diaphragms on the DPA omnis are made of stainless steel or nickel. These materials can be tightened very hard and are less compliant.

Finally, remember microphones are an analog device and have a sound/tone of their own even though the goal in design is a pure reproduction of the original sound. After years of use, a microphone will wear out and will change its tone.

## Sound Design and Microphone Applications

Sound design for television goes beyond the faithful reproduction of a sound source. Interesting sound design should enhance the viewing experience for the audience. The first question is whether the production is live or if the sound track will be produced later. Beautiful sound tracks are sculpted every day in recording studios, but the sound capturing and recording is approached differently for a production that will have the audio mixed and produced after the event. The sound designer, audio technician and producer will identify the critical sound elements that must be recorded on location and insure the audio is properly recorded for editing.

Sound design for a live event entails capturing the necessary sound sources, organizing the sound sources for a mix and combining and blending the audio for various output formats. It takes a skilled sound designer and mixer to capture and bring to life a live television production. A live sound track must have all the contributing sound sources, announcer(s), atmosphere and specific microphones plus recorded music and effects literally at the fingertips of the sound mixer.

**Figure 7.22** Super-directional DSP microphone AT895.

Sound design for any broadcast event consists of four basic layers: announce, atmosphere, event-specific sounds and playback sources. Event-specific sound is very diverse and specialized, ranging from stage miking for a music ensemble to sideline microphones for football. Announcers can range from podium microphones, wireless body packs to sportscaster's headsets and lavalieres! Recreating a proper atmosphere is key to the foundation for a stereo and surround mix.

Television is a visual medium and by definition the audio has to contribute to the pictures. First, examine all camera angles and the variation in content of each camera. Microphone placement is dependent on camera perspective and production values. Sports television coverage is capable of putting the viewer in the stands or in the car and undoubtedly provides a depth of coverage that is unattainable to the ordinary spectator in the stands.

**Figure 7.23** Wireless handheld camera with stereo microphone. Microwave transmitters for cameras should have two channels or more of audio for handheld cameras.

The desire is to capture the atmosphere of the event where basic coverage begins from the spectator's perspective or audience's point of view of the event. Do not depend on the camera or other close microphones to provide an adequate atmosphere base for the mix. Microphones should be specifically placed to capture the atmosphere and microphones must be considered for surround-sound miking. Atmosphere is not strictly limited to audience. The atmosphere at a race track is the cars and at the golf course the atmosphere is nature (hence the name "birdie" microphone). Peter Adams used three different microphone positions for the in-field "roar" microphones at Talladega Motor Speedway to create atmosphere sound.

Television provides many different perspectives with the change of the camera, and most sports productions are constantly changing perspective. Car racing is a sport that demonstrates the huge changes in camera views, from ultra wide, to in the pits, to in the car, and then to various camera positions around the track. Each view has a distinct soundscape denoting depth of field and spatial imaging. Tight audio-follow video is critical to punctuate and enhance the camera cut! For example, when the director cuts to the "in-car camera" the sound changes quickly to inside the car with all outside microphones out of the mix. The sound from the driver's perspective allows the viewers to hear the engine throttling up and down and is not available to the viewer in the stand.

A camera that adds a dimension of speed is a camera mounted on the track with a head-on view of the cars, known as the speedshot camera. At FOX Sports, Fred Aldous has a specific combination of microphones for cameras beside the track. There is a microphone up track away from the camera and a microphone down track of the camera, and a stereo microphone above and away from the camera. As previously stated, with close microphone techniques, small changes in the position of the microphones will result in changes of the sound source in the sound field. This is particularly evident with the speedshot camera because of the speed of the race car.

There is also a timing relationship with the sound to the picture that determines where the approach microphone is placed in front of the camera. If there is just one microphone on the camera, then the image of the car will be past by the time the sound gets to the speakers. The distance the microphone is moved away from the camera adjusts to where in the picture frame, from left to right, the full intensity of sound is heard. If the microphone is too far up track from the speedshot camera, the sound envelope will peak before the race car fills the picture frame. That distance will determine the full intensity of the sound of the car to the size of the car in the picture frame.

The combination of mono microphones up and down track and a stereo microphone at the camera gives a very strong left-speaker approach and a strong right-speaker depart with a stereo base. Fred also constructs a surround-sound mix that orients the sound of the speedshot camera beginning in the front left passing through to the rear right speaker, giving the illusion the cars are coming across the viewer.

Even sports like football and basketball that are covered from cameras above the field of play requires that the sound perspective change when the director cuts to the handheld cameras on the field or sidelines or other closer perspectives. Tennis is an unusual sport because there is a camera angle over the near-court player. This changes the perspective and is a problem for a stereo mix. However, the surround sound view can incorporate the front left and rear left speakers to enhance the camera perspective.

**Figure 7.24** Large diaphragm microphone in the diffused sound field for surround sound.

Often music and entertainment television productions are in a venue with good acoustics that can contribute to the reproduction of the event. In this situation the audio producer will probably want to "capture the acoustic and spatial atmosphere" of the venue. Music production usually has a very strong front stereo music mix using closely placed microphones on the instruments. The audience sound perspective of the room and stage usually will not change with the camera cut, except for a special effect. It is very common in music productions to show the audience from the musician's perspective. This point of view (POV) would be a 180-degree reversal of the front cameras and to reverse the sound field would be disconcerting to the viewer (Figure 7.25)!

**Figure 7.25** The mix of the microphones around the drums will not change perspective when the visual changes the perspective of the drummer.

To achieve a desirable balance and variety in the sound mix, a combination of stereo and mono shotguns, "close-miking" boundary, lavaliere and specialty microphones should be utilized. The audio mixer then produces and blends the spatial atmosphere and specific event microphones to achieve a pleasant soundtrack that is appropriate to the visuals.

Directional and omnidirectional microphones are available as dynamic or condenser and in a wide variety of sizes. Certainly size and housing will determine the choice of microphones because, as previously stated, placement of the microphone is critical for good sound, but safety is also a major concern. For example, in gymnastics the athlete does not wear shoes and has no protection or padding in the event of a fall. Shotgun microphones were the tool of choice for most television sports productions through the 1980s and had to be mounted on stands away from the action. Specialty microphones like small-diaphragm cardioid and low-profile surface-mount microphones provided good solutions to distant shotgun microphones and should be considered when designing a microphone plan.

Miniature microphones have a legacy as a lapel microphone, usually worn on the clothing at mid-chest level. Miniature microphones were often overlooked because their design was usually omnidirectional and the contemporary thinking was to use high-quality shotgun microphones. No matter how good the design or who the manufacturer is, shotgun microphones will affect the tonal characteristics of a sound, where a good omnidirectional microphone should not. Separation is a good thing, but listen to the microphones and if the leakage from one sound source appears in another microphone and it sounds natural, then this can be beneficial because it adds natural tone and character of the sound source.

**Figure 7.26** Miniature cardioid microphone on gymnastics landing mat. This microphone offers the advantages of a clip-on flexible mounting arm and windscreen.

## Surface-mount Microphone

Often it is desirable to place a microphone on a flat surface or tabletop close to a sound source. This presents a problem because of the short delay time between the direct and reflective sounds creating an out-of-phase delay known as the *comb filtering* effect. The boundary microphone, which is a type of surface-mount microphone, solves this problem because the capsule is so close to the reflective surface that there is no delay time or phase cancellation. A boundary surface reflects the sound waves back into the microphone capsule, resulting in an increase in gain and a perceptible increase in clarity.

**Figure 7.27** Boundary microphone on hockey glass.

 The boundary microphone is specially constructed with the microphone element flush against a flat barrier. The diaphragm is made part of the larger surface and benefits from the surface pressure build-up, which is even across most frequencies (Figure 7.28).

The pressure zone microphone (PZM) is another type of surface-mount microphone that has a distinctly different approach to sound capture and design. It has an omnidirectional capsule that "looks down" at the boundary, while the phase coherent cardioid (PCC) microphone has a supercardioid capsule that "looks across" the boundary.

**Figure 7.28** Close-up of boundary microphone housing capsule and electronics.

The PZM was first introduced as a method of recording "PRP - Pressure Record Process" by Ed Long and Ron Wickersham, with the microphone being manufactured by 1980 by Crown, the amplifier company. (See US Patent 4,361,736.) By placing the capsule of the microphone in the "pressure zone," there is the coherent summation of all the direct and reflected or delayed sound, resulting in an additional 6 dB of acoustical gain. The microphone capsule faces downward, which eliminates any direct sound and problems associated with on-axis reinforcement or peaking in the sound response due to the dimension of the capsule and size of wave length. This design also reduces local diffraction properties, which are common with free field microphones.

The PZM's polar pattern is hemispherical when mounted on the proper size boundary and is inherently flat both on- and off-axis and more directional than a cardioid microphone. Size of the boundary determines frequency response. A 2 ft by 2 ft boundary has a frequency response down to 80 Hz. You can control the directionality of a PZM by adding a second and third boundary, which also gives additional gain. The second way to control the pickup pattern is by acoustic absorption.

The Crown PZM microphone was conceived by the designers as a recording process, but because of its size and housing design the microphone had some interesting possibilities! Motor sports presented a problem because there are camera positions (roof camera) that cover the pits and the camera microphones do not have any close direct sound. ABC Sports used PZM microphones on the pit walls at the Indianapolis 500 and had direct sound from every pit for every camera view. The downside is the cost! This microphone plan required an additional effects mix, mixing console, microphones, cabling, power supply, line amplifier and audio assist to install it! This is not insignificant.

## Microphone Phasing

*The table-top example is an extreme example of how every hard surface or enclosed environment creates a unique sonic characteristic, because sound is reflected and absorbed from surfaces. When a microphone captures the sound it is a mixture of direct sound and reverberant sound and the net result is that the sound information going into the microphone diaphragm has its various time delays, resulting in colored information.*

*Acoustic phase interference may also occur when only a single microphone is in use. This happens when sound is reflected off a nearby surface and arrives at the microphone slightly after the direct sound. The adding together of the two signals may give problems similar to those encountered in improper multimicrophone set-ups. (The phase interference will be most noticeable when the reflected sound arrives at a sound pressure level that is within 9 dB of the direct sound.)*

*There are several ways to eliminate this problem. First, try putting the microphone closer to the sound source. Second, move the microphone farther from the reflective surface. Third, use a microphone specially configured to be placed extremely close to the reflective plane. When using a low-profile directional Audio-Technica boundary or "plate" microphone, for example, the microphone capsule is so close to the surface that the direct sound and the reflected sound arrive simultaneously and add together rather than cancel. This technique can prove very helpful on the apron of a stage, on a table or desk for conference use, or on the altar of a church.*

*When two or more microphones are to be used in close proximity, phasing between microphones and reflective surfaces are a critical consideration. Phase adversely affected signal levels and tonal balance abruptly with small movements of the sound source or the microphones. When phasing occurs with stereo microphones there may be poor imaging, imprecise location of instruments, and reduction of bass.*

*Sound from a single source arrives at two different microphones at different times. A microphone that is further from the sound source will pick up the source with a delay in relation to the first microphone. When you combine the close and distant mic signals in your mixer, certain frequencies cancel out due to phase interference, creating what is know as a comb-filter effect.*

*Audible comb filtering can occur whenever two or more mics pick up the same sound source at about the same level but at different distances, and are mixed to the same channel. The frequency response of a comb filter has a series of peaks and dips. This response often gives a thin, hollow, filtered tone quality.*

*In general, place mics close to their sources and keep the mics far apart to prevent audible comb filtering. This problem can be minimized or eliminated by following the 3-to-1 rule. Separate microphones by at least three times the microphone-to-source distance. This creates a level difference of at least 9 dB between microphones, which reduces the comb-filter dips to an inaudible 1 dB or less.*

*Remote broadcasts rely on copper wire and three-pin XLR connectors for the basics of analog signal interface. Complete phase cancellation of 180 degrees can occur if pin 2 and pin 3 of the connectors are wired out-of-phase to each other.*

## Shotgun Microphones

Sports audio is a challenge because the A-1 is always fighting to isolate specific sounds in a very noisy environment. Microphone directivity has been the Holy Grail in microphone design but is elusive because of the omnidirectional and reflective characteristics of sound waves. Shotgun microphones are effective when used to track a moving sound source and if the microphone is properly mounted to a camera or pointed by an operator at the moving sound source. Cameras with large lenses should have mono shotguns to try to capture the sound source from greater distances. There is a point of diminishing returns where a camera is so far into the diffused sound field that it does not contribute to the mix. The further the camera position, the more diffused the sound and the less likely to need a shotgun microphone. However, a distant camera may give a good atmosphere base and perspective of the venue and a stereo microphone may be appropriate for this location.

**Figure 7.29** Wind Zeppelin with special clamp to mount the microphone away from camera.

A mono shotgun microphone should be aimed at the moving sound source by the camera or microphone operator. Shotguns and other highly directional transducers have a "sweet spot" where sound capture is optimal, but once the sound source moves through the sweet spot, the on-axis sound becomes off-axis and loses its presence in the mix.

Shotgun microphones are effective because their design utilizes off-axis cancellation characteristics but, remember, this phase cancellation has an effect on the tonal characteristics of sound. It was customary to use shotgun microphones on the speedshot camera; however, this has been changed to an omnidirectional microphone because it can take more SPL without overloading and the reflective sound off the track wall adds to the sonic characteristics of the shot.

### Stereo/Surround

Stereo microphone techniques are essential not only for stereo sound capturing and recording but for surround sound as well. There are basic stereo microphone techniques that derive stereo information from variations in sound intensity and phase. These techniques are *coincident, mid-side* and *spaced pairs*. Coincident microphone techniques were pioneered in the 1930s by engineer Alan Blumlein. Blumlein's array produced intensity-related information for stereo recording. These principles are incorporated into modern mid-side techniques. The mid-side (M/S) technique uses a forward-facing cardioid and a side-facing figure-eight pattern. The outputs are matrixed to output an L+R signal and an L–R signal. (More details on MS techniques will follow later in this section.)

Coincident is a generic term for placing two microphone capsules oriented between 60 and 90 degrees of each other in the same vertical plane. The coincident technique usually employs a pair of directional microphones aiming at the left and right sides of the sound source with some overlapping. Each microphone signal is assigned completely to either the left or right channel, respectively. This creates a stereo image because of the difference in sound pressure intensity and time differences between the two microphones as the sound wave passes through them. The time differences cause phasing between the capsules, which contributes to the effect.

Each microphone has an on-axis sound pattern with some overlapping, which gives a strong center image. The "off-axis" attenuation and quality of the polar pattern determines the width of the stereo spread. Figure 7.30 shows two shotgun microphones with the capsules symmetrically angled and on top of each other. XY pairs can be employed very near the sound source and still provide spatial separation and audio cues to the ears and brain.

Coincident and near-coincident microphone techniques work well with larger sound fields like football stadiums or for the up cameras on NASCAR coverage. Coincident microphone techniques with shotgun microphones or other directional microphones give a tighter on-axis sound at greater distances from the sound source. Coincident techniques provide good sound-field location, along with good localization, and listeners will perceive a depth to the space.

**Figure 7.30** Two short shotgun microphones on top of camera lens used in a near-coincident pattern.

Numerous recordings and documents have been published on "near-coincident techniques" by the French and Dutch radio organizations. This technique was pioneered by ORTF, the French National Broadcasting Organization, which uses two cardioid mics angled out 110 degrees from the center line with capsule spacing of 17 cm. Two cardioid mics angled out 90 degrees from the center line with a capsule spacing of 30 cm is the standard approved by the Dutch Broadcasting Organization (Nederlandsche Omroep Stichting).

## Stereo Microphones

To consolidate the microphone package, contemporary microphone designers utilized a coincident and a mid-side approach to achieve a stereo image, and then packaged the two capsules in the appropriate housing.

XY capsule orientation is incorporated into two of Audio Technica's microphone designs. The AT825 (Figure 7.31) is roughly the size of a hand microphone with a larger diaphragm housing for the two microphone capsules. This microphone has been used extensively in sound-effect recordings but is also used often to provide a stereo atmosphere microphone.

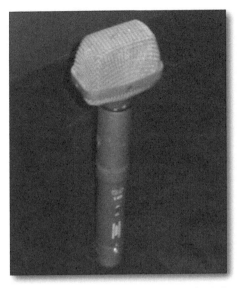

**Figure 7.31** XY capsule placement is incorporated into this Audio Technica AT825.

The stereo boundary microphone is useful to maintain a stereo perspective while close miking the sound source. (See Figures 7.32 and 7.33.)

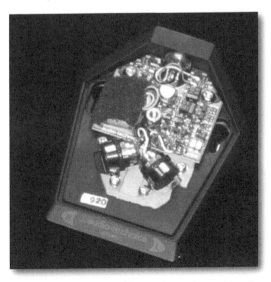

**Figure 7.32** XY capsule orientation is incorporated into a boundary microphone AT849.

**Figure 7.33** Stereo boundary microphone mounted on wall. POV sound for robotics camera backstage.

Shotgun microphone designers utilized a mid-side approach to achieve a stereo image and then packaged the two capsules in a shotgun interference-tube configuration. Stereo shotguns use the microphone capsule orientation, housing design and phase techniques to create a spatial image. As mentioned, MS is a variation of the coincident miking technique with overlapping pickup patterns. The MS technique uses two different types of microphones. The "mid" microphone is a directional cardioid or hypercardioid aimed directly at the sound source. The "side" microphone is a bidirectional (figure eight) microphone angled 90 degrees off-axis from the mid microphone and facing the left and right sides of the sound field. This approach passes the audio through a sum and difference matrix, which uses a phase processing to extract a stereo image from two microphone capsules. (Refer to Figure 7.34.)

**Figure 7.34** Stereo shotgun microphones are used extensively on handheld cameras.

The MS matrix takes the sum information M + S, and sends it to the left channel, and the difference information M − S, and sends it to the right channel. What gives MS stereo its flexibility and mono compatibility is the sum and difference matrix. When the left and right signals are combined, (M + S) + (M − S) = 2M, the sum is M information only. If the MS signals are recorded discretely (without going through the matrix) the signals can be matrixed in post, and the relative gain of the two signals can be adjusted to provide varying degrees of stereo width. The MS approach offers 100% mono compatibility and the ability to alter the sound field in post production.

There are some general guidelines about camera-mounted microphones. All handheld cameras will have a stereo microphone. If the camera has only one audio input or the camera is older than ten years, then the microphone signal will come to the truck by audio cable. Both channels will come back on audio cable, not one down the camera electronics and one down audio cable. All cameras on field-of-play will have a stereo microphone on them. Typically this will include low cameras at track and field or tennis, where the camera is on wheels. Generally, the further the camera is away from the field-of-play, the less likely that the camera will have a microphone. Finally, use a proper mount and get the microphone off the camera lens.

### Useful Tricks with Mid-side Stereo

*When recording voice tracks, the sound recordist and boom operator need to be very conscious of background noise. MS stereo recording techniques can be used to record dialog in large reverberant spaces. Mono microphones often result in objectionable amounts of room sound versus program material. When MS recordings are summed, side information (most of the reverberant signal) is cancelled, not added. MS recording, compared with other stereo recording techniques, provides the best method for film and video sound.*

*The biggest criticism of stereo microphones is that there is not enough forward reach! Stereo with more reach can be created with a hypercardioid microphone and a figure-eight microphone positioned appropriately. A longer tube hyper microphone can be paired with a separate figure-eight microphone.*

*Single-tube stereo microphones are convenient and easy to use with handheld cameras and microphones from Audio Technica, which can output XY and MS audio. The side fill can be varied as desired. The MS approach has been used in single-tube shotgun packages from Audio Technica and Sennheiser*

Another microphone technique, known as a *spaced pair*, employs two microphones of the same polar pattern set symmetrically along a line perpendicular to the sound source. A spaced approach derives a stereo image from intensity differences plus the time delay between the two microphone capsules. Spaced omnidirectional microphones can result in an extremely open and lush sound field; the downside is possible phase anomalies present when the signal is summed to mono. Spaced techniques also require critical microphone placement for accurate results. Figures 7.35, 7.36 and 7.37 show examples of spaced-pair microphones.

**Figure 7.35** One microphone of a spaced pair of omni microphones.

**Figure 7.36** The microphone on the "robotic camera" is the second microphone of the "spaced pair."

**Figure 7.37** A stereo microphone (AT825) is mounted above the camera to provide a strong left/right orientation.

The Decca Tree is a variation of spaced-pair techniques that originates from stereo recording at Decca Recording Studio in London. Stereo was emerging and engineers placed three microphones on a bar, one on the left, center and right. The left and right microphones were assigned respectively, and the center microphone was fed to both channels, although 3 dB less than the other signals. This technique has applications in surround sound.

## Surround Sound Microphone Techniques

The stereo and surround-sound image is derived from mono microphones that are properly mounted and then combined and processed to create a dimensional sound field. Spatial orientation is a subjective and a creative aspect of sound mixing. The sound designer and mixer control the spatial orientation of the sound mix with good microphone placement and panorama orientation in the mixing process.

There are basically two different approaches to capturing a surround-sound image—a *fixed array* or a *tuned array*. A fixed array uses permanently placed microphones arranged at fixed distances on a multi-arm or tree-like stand. A spaced tree uses separate microphones in each position, allowing the designer to change microphones, patterns and space between microphones. Remember, spatial imaging can be accomplished through amplitude and delay between combinations of microphones and the sound source.

Engineers have made recordings using a fixed-array design in the shape of a human head with microphones in each ear position. A similar approach used for surround has multiple microphone capsules mounted in an elliptical-shaped configuration. Arrays such as the Soundfield and Holophone microphones utilize a rigid housing with transducers arranged in the shape (Figure 7.38). The Soundfield microphone is a four-capsule array that processes the audio to change the spatial characteristics of the sound capture. The microphone is capable of changing pickup patterns and with four capsules can produce 5.1, 6.1 and 7.1 surround.

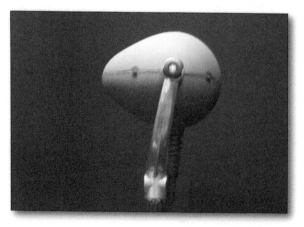

**Figure 7.38** The Holophone is a fixed-array surround-sound microphone.

The original Holophone had discrete outputs and the power-supply box could generate a stereo matrix signal of the seven capsules. The biggest advantage is the packaging of the seven different microphone capsules, which makes it easier to rig and mount for a production. At the Grammy Awards a single Holophone was suspended above the audience about 30 feet from the stage. The only drawback is not being able to move the capsules.

A tuned array resolves the limitations of not being able to move the capsules. With a fixed array the sound designer has to find a generally good area to place the fixed array microphone. One of the problems is with the low frequency capture known as LFE. You have to find an LFE sound source or you will end up using bass enhancers or synthesizers to derive an LFE. With the tuneable array, the sound engineer places a specific microphone to capture the LFE.

In this sense the venue and coverage will dictate microphone placement. Front left and right and surround left and right are in essence spaced pairs. Beginning at the 2000 Olympic Games, AT4050 large-diaphragm microphones were used for surround channels. The spaciousness is derived from intensity and time differences between the microphones. The tuned array tends to be very phase coherent because the microphone is in the diffused acoustics sound field and not in any of the primary stereo images. Additionally, a properly positioned tuned array will maintain the sound image through the various matrixed encoded transmission and distribution methods.

Even with the purest of stereo and sound reproduction, an array of various combinations of microphones is used. Mono spot microphones are used to supplement and balance the sound. The mixture of microphones and sounds can now be placed in a variety of combinations and positions in the sound field to create a surround, stereo and mono sound mix.

## Microphone Protection and Mounting

Microphones are particularly susceptible to any wind movement over the microphone capsule. Windscreens are usually made of a foam material that does not affect the sound wave but dissipates the wind velocity as it travels through the foam. Windscreens are an essential accessory that should be used indoors as well as outdoors.

Foam wind screens are a basic precaution, but if the event is outdoors then additional weather protection such as Wind Zeppelins with fuzzy covers may be necessary. Wind Zeppelins completely enclose the microphone in a cage that looks like an airship. With the addition of a fuzzy windsock, wind velocities of up to 40 mph can be withstood.

**Figure 7.39** Zeppelin-style windshield with grip handle. The handle is threaded in the bottom so it can be screwed onto a microphone or other type of stand.

Microphones are prone to physical vibrations transferred through a rigid mounting. Some handheld microphones are designed to withstand some handling noise but shock mounts should be used not only for good mounting but to isolate the capsule from extraneous vibrations.

In extreme conditions it may be necessary to use the foam shield, the zeppelin and fuzzy (Figure 7.40). Outdoor sports will continue in wet weather and in rainy conditions condoms are used to enclose the microphones and keep moisture out. Note—use only nonlubricated condoms!

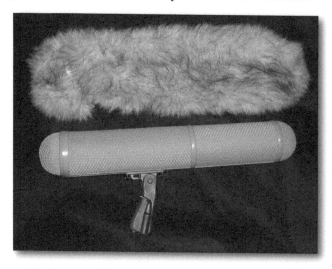

**Figure 7.40** Rycote zeppelin, fuzzy and pistol grip.

# Specialty Microphones

On- and off-camera announcers have special needs and the manufacturers have designed several announce microphones and headset configurations for this purpose. Announcers used a variety of stand-mounted microphones till Sennheiser standardized the "headset boom microphone." In the early 1980s, CBS modified a Sony headset because of its look and comfort and Sennheiser again standardized a production model. Field production for sports usually requires the announcer to wear a headset with a boom-mounted microphone (Figure 7.41).

**Figure 7.41** Headset microphones with boom arm for microphone.

The following are some common specialty microphone types and their descriptions.

*Handheld Microphones.* A microphone held by an entertainer may endure rough handling and extreme shock and improper use (Figure 7.42).

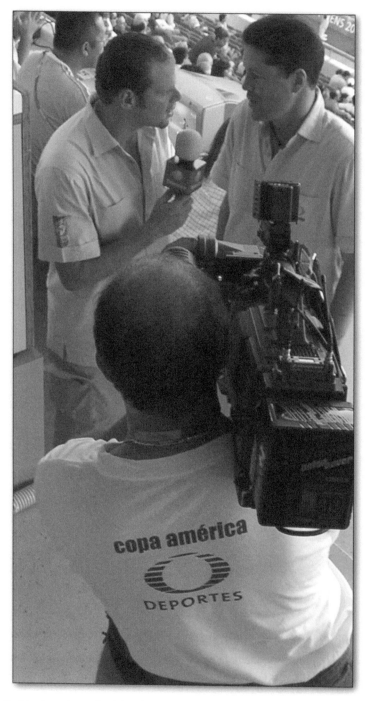

**Figure 7.42** Handheld microphone for interview on sidelines.

*Coles lip microphone.* A British company developed for the BBC a "lip microphone" that literally touched the announcer's lips. The lip microphone is a ribbon microphone and still in use today by the BBC. By shortening the distance from the announcer's mouth to the microphone, a much stronger voice signal helps mask the unwanted crowd.

*Podium microphones.* Television productions such as award shows and game shows use microphones on a boom extension and mounted to a tabletop or podium.

*Contact Microphone.* The contact microphone is useful because it does not pick up acoustically transmitted sound vibrations. The contact microphone is attached to a rigid surface that acts as a sound board just like a piano. The contact detects the vibrations in the sound board and converts them to an electrical impression.

When you attach it to a resonate surface, the effective coverage extends proportionally to the size of the surface. Figure 7.43 shows the A2 digging a hole in the landing pit at high and triple jump. This is a particularly good application for this rig because the hole is one meter deep and the sand is kept wet. The sand pit is raked after each land and the microphone has to be deep enough to insure that it does not interfere with the competition. The contact microphone is completely sealed and water proof. Figures 7.44a and b show the contact tightly fixed to a plexiglass sound board. A wood surface is an excellent sound board and the contact microphone has possibilities attaching it to a Veledrome track, basketball court, bowling alley, boat hull and even a race car frame.

**Figure 7.43** Digging out the landing pit at long jump.

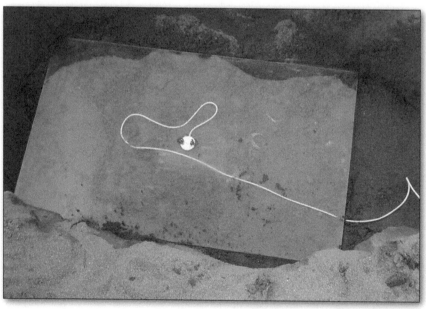

**Figure 7.44 (a)** Contact microphone mounted on Plexiglass. **(b)** Closeup of contact microphone.

*Parabolic microphone.* Parabolic microphones use a mathematically formulated bowl to reflect sound to a focal point. A microphone is mounted at the focal point of the reflector and is focused by moving the capsule in and out from the reflector to obtain a highly directional response. The greatest gains are when the reflector is large in size compared to the wavelength. When a 3-foot parab is in focus, it is effective in the high mid-frequencies around 8,000 Hz. Parabolic receptors are used primarily in American football and variations of this concept have been used in surveillance. (See Figure 7.45.)

**Figure 7.45**  Parabolic microphone.

*Digital processor microphone.* Audio Technica developed a digital processor microphone that is capable of significantly rejecting certain unwanted frequencies. The AT895 has the ultra-directivity like the parab at certain frequencies, making it useful for specified applications. The microphone has been used very successfully in World Cup and Olympic football since its delivery in 2000 (Figures 7.46 and 7.47).

**Figure 7.46** AT895 digital processor (DSP) microphone.

**Figure 7.47** The AT895 uses five microphone capsules to capture the analog audio for digital processing.

# Room Acoustics—Direct Sound, Direct Reflections and Reverberant Field

When the production is outdoors, issues arise with wind noise and disturbances like traffic or airplanes, but there usually are not acoustic problems from reflective surfaces. Square indoor venues like a basketball venue suffers from poor room acoustics that will have an impact on sound coverage and design. Standing waves created by bass frequencies particularly have an effect on shotgun microphones (pressure gradient).

Room acoustics can have a significant impact on the quality of the location audio. Parallel glass walls in stadium press boxes are normal and cause echoes when the announcer is talking. Sports production uses a dynamic or condenser microphone on a boom very close to the announcer's mouth. While this significantly increases the direct signal over the reflected signal, the reflected signal still reflects back into the microphone. Some basic room treatment should be applied to absorb some of the sound reflections.

# Getting the Microphone Signal to the OB Van

The signal path back to the television truck can be a long and convoluted path. Quality cable and consideration of the cable path will help minimize any outside interference to your cabling. Microphones generate extremely small amounts of AC voltage, which makes the audio path very prone to interference. Microphone signals are so low that they have to be significantly amplified and that also amplifies any defects in the cable paths. Microphone connectors or other metal parts that touch an earth ground can generate hum. Water can easily interfere with the electrical connection and diminish the quality and level. Water shorting across audio connectors will introduce crackle and noise.

In large set-ups there can be multiple television production vans and stage lighting requiring high-voltage power runs everywhere. Parallel paths along high voltage will potentially introduce induction hum into your audio. Lighting and dimmers can also cause interference to the audio signal. Use shielded and balanced audio cables which, by design, provide a certain amount of common mode noise rejection.

Microphone connections increasingly use fiber systems, which minimize many interference problems, plus fiber is very lightweight, making it possible to have longer distances between connectors. However, multipair bundles of copper wire with XLR connectors are always going to be used in the installation. See Chapter 5, "Putting It Together."

Cameras normally have two channels of audio that are modulated with the other camera signal down fiber or triax to a camera control unit (CCU) in the OB van. The quality of the audio path through cameras has significantly improved and most cameras provide phantom power and an "input pad." In extremely loud situations you will need to insert an additional pad of up to 25 dB between the microphone and the camera XLR input.

Most audio manufacturers wire their microphones to conform to the most popular industry convention: positive acoustic pressure on the diaphragm generates a positive voltage on Pin 2 of the 3-pin output connector. Of course, consistent phasing (polarity) must be preserved in all of the cables between the microphone(s) and the electronics. The term "out-of-phase" is often used to describe a microphone that is wired with its polarity reversed with respect to another.

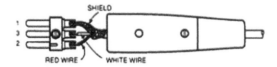

**Figure 7.48** Balanced Audio XLR.

## *Balanced Output*

Most microphones offer balanced output. A balanced output offers real advantages to the serious record-ist. Balanced lines are much less susceptible to RFI (radio frequency interference) and the pickup of the other electrical noise and hum. In a balanced line, the shield of the cable is connected to ground, and the audio signal appears across the two inner wires, which are not connected to ground. Because signal currents are flowing in opposite directions at any given moment in the pair of signal wires, noise which is common to both is effectively cancelled out (common mode rejection).

This cancellation can't occur when only one signal wire plus the shield is used. Of course, it is possible to wire a low-impedance microphone directly to an unbalanced low-impedance input, but the noise-canceling benefit will be lost. This should not be a problem with short cable runs, but if longer cables are used, a balanced input is preferred.

# Summary

Sound design for television is a process of listening, visualizing and testing. Listen to the environment! What sounds are an essential part of the soundscape? What sounds are natural and what sounds are naturally enhanced? The atmosphere of an event may be comprised of many elements blended together to provide the spatial orientation for a stereo and surround mix. The sound designer knows the soundtrack or effect that they want to hear and proper microphone selection and placement is the key to capturing it.

As I have emphasized throughout, sound design is a creative science and microphones are the basic tools for sound capture. You have to understand the science of microphones to make intelligent choices but experiment with new designs and share ideas. The size and performance of microphones have contributed to a variety of approaches to the art of sound.

There are many opinions on how things should be done, but there is no set way of doing things with microphones! A lot of sound designers relate mixing to painting and sometimes it is the subtleties of a color or sound that works better than a wall of sound. There is room for a nonconventional approach to sound capture and delivery, as long as it sounds good! Learn to trust your ears and understand what makes up a good soundtrack (Figure 7.49).

The sound designer must test the design, because sometimes you do not know how your tools are going to act in a particular application. A wonderful design that cannot be implemented is a bad design, and microphone placement and protection go along with microphone selection.

**Figure 7.49** Innovation—Fred Aldous had experience with car racing but took a completely fresh approach to microphone selection and placement when he designed the sound for the NASCAR coverage on FOX. NASCAR is an extreme in microphone selection and placement and is even further complicated with surround sound.

## *References*

Bartlett, Bruce, "Microphone Placement for Stereo Recording," *Radio World*

Streicher, Ron and Dooley, Wes, "Techniques for Stereophonic TV and Movie Sound," *AES*, 1988

www.audio-technica.com

www.dpamicrophones.com

www.sennheiser.com and www.sennheiserusa.com

www.shure.com

# 8 Wireless Systems

A large-scale television broadcast requires wireless PLs, IFBs, microwave links, wireless cameras, handheld radios, and wireless microphone systems, along with the usable frequencies to operate on. Wireless offers a degree of freedom not available with wired microphones and certainly has changed the broadcasting of major sporting events such as motor sports and football.

There are many situations where the use of a wired microphone is not desirable or is absolutely impractical. For years, audio dealt with wireless microphones breaking up on-air and always had a hard-wired microphone installed somewhere as a back-up. As technology and experience progressed, wireless transceivers have proven their reliability and can duplicate the performance of a wired microphone without coloration or distortion. For the audio technician, this migration to more and more wireless offers new opportunity and skill requirements that did not exist before the early 1980s.

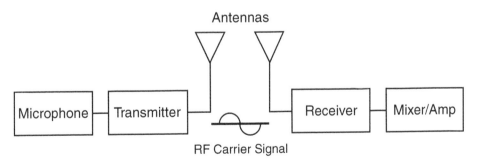

**Figure 8.1** A basic wireless system consists of a transmitter, pair of antennas and a receiver.

Wireless microphones are found on pit and sideline reporters, and in bases on players and coaches. The wireless microphone has significantly put the reporter close to the action with unparalleled freedom.

A wireless microphone system consists of a microphone, a miniature radio transmitter, radio receiver and antenna (Figure 8.1). The transmitter operates like a tiny version of an FM radio station. Similarly, the wireless receiver is like a home FM tuner or car FM radio. The transmitter converts the audio signal into a frequency-modulated carrier (FM), and then radiates the modulated RF carrier where it is picked up by the receiver.

# Transmitters

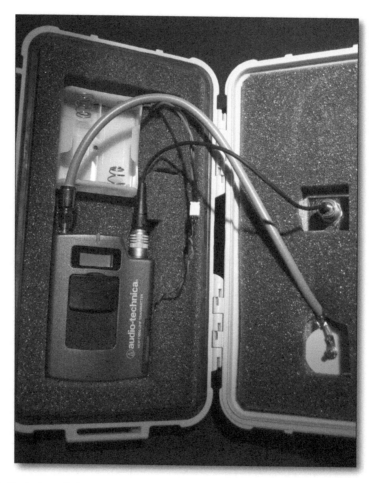

**Figure 8.2** Audio Technica wireless transmitter housed in a waterproof case used at sailing. C-size batteries are used to lengthen the transmitter life to over eight hours. The antenna and microphone are mounted on the external surface.

**Figure 8.3** Wireless transmitter in jump station of equestrian event.

There are three general types of wireless microphone transmitters: handheld, body-pack transmitters and plug-in type transmitters. Handheld transmitters include the microphone element, radio transmitter and battery in one package. Bodypack transmitters are housed in small, thin, flat, rectangular packages that are intended to go into a pocket, worn on the belt or be concealed on the body. This package includes the transmitter, battery and a small connector to attach an external microphone (Figure 8.4). Plug-in transmitters are designed to be inserted into the connector of a standard dynamic microphone. This type of transmitter is generally popular with television ENG crews working in the field.

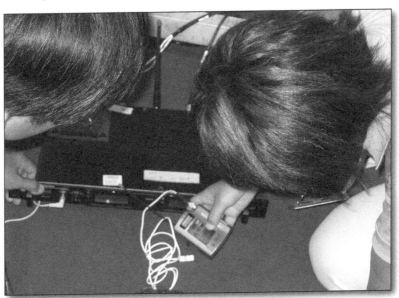

**Figure 8.4** Bodypack transmitters are attached to people and to equipment.

Bodypack transmitters are generally used with an external microphone, either a lapel, headworn, or instrument pick-up. Professional transmitters provide phantom power for most electret devices directly from the internal batteries. Some transmitters have special circuitry to accommodate guitar pickups and other types of special-purpose pickups and they usually provide considerably better performance than transmitters that do not have such circuits.

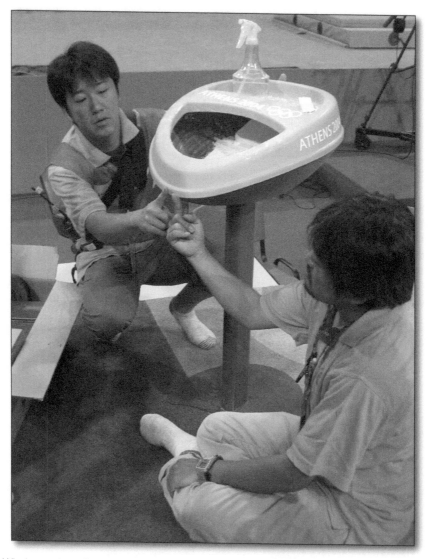

**Figure 8.5** Wireless microphones are increasingly used in sports where a wired microphone would interfere. For example, the chalking stand in gymnastics can be moved around by the coaches. The clapping and hand rubbing may be shot on a camera further away, not capturing the sound. A wireless microphone permits perfect microphone placement and freedom of movement.

Handheld transmitters are a combination of microphone element and transmitter built into the same package (Figure 8.6). These devices are popular for vocalists, and speakers that prefer to hold on to the microphone. Handhelds generally are available with a variety of different microphone capsules in electret, condenser and dynamic elements.

**Figure 8.6** Audio Technica handheld true wireless condenser microphones. Wireless microphones speed up scene changes because there is no cable to hook up or screw up.

Handheld transmitters are usually the best choice for vocal performances, live interviews and situations where the transmitter is passed from person to person. Bodypack transmitters are appropriate for most other applications.

**Figure 8.7** Wireless engineers walking through the "hot-zone" where flawless transmission and reception are necessary.

A common problem with transmitters is the location and adjustment access of operational and function switches and control pots. Switches or pads that are too easily moved or could accidentally be turned off during a performance will affect the popularity of a transmitter. Many adjustments are screen menu-driven and settings can be locked.

Setting the proper input gain is the most important adjustment on a wireless transmitter, requiring proper access to level adjustment. Suitable setting of the input gain is critical to the signal-to-noise ratio of the system. If set too high, severe distortion and/or compression of the dynamic range will occur. If the gain setting is too low, the desired audio signal may be too close to the noise floor and not usable.

## Radio Transmission

A wireless transmitter emits radio waves in a series of electromagnetic variations or modulations.

If the amplitude of the wave is varied it is known as amplitude modulation or AM. If the frequency is varied or modulated, it is called frequency modulation or FM. The audio signal is modulated onto a radio carrier. More audio information can be sent in the typical FM signal, allowing higher fidelity audio signals to be transmitted. Additionally, FM receivers are inherently less sensitive to many common sources of radio noise, such as lightning and electrical power equipment, because the AM component of such interference is rejected.

The radio wave itself is the carrier of the information and the information is actually contained in the amplitude variation or frequency variation of the radio wave. The radio wave for an FM signal has constant amplitude, which is determined by transmitter power and fundamental transmission frequency. The basic radio frequency is varied up and down (modulated) by the audio signal to create the corresponding radio signal. This frequency modulation is called *deviation* since it causes the carrier to deviate up and down from its basic or unmodulated frequency.

**Figure 8.8** Wireless bodypacks.

**Figure 8.9** Transmitter plugs directly to microphone.

Radio waves are quantified by their frequency in the radio spectrum. The range of frequencies extends from a few Hertz through the Kilohertz (kHz) and Megahertz (MHz) ranges to beyond the Gigahertz (GHz) range. Frequency is correlated to the wavelength. The higher or greater the frequency, the shorter or smaller the wavelength. A wavelength is the physical distance between the start of one cycle and the start of the next cycle.

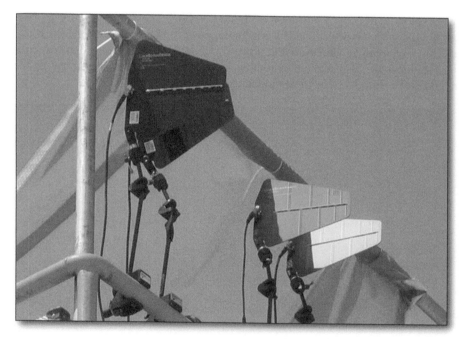

**Figure 8.10** An antenna tower where several Log Periodic Dipole Array antennas sometimes called "paddles" are mounted.

The strength or magnitude of a radio wave is known as the amplitude of the wave. Variation in the magnitude of the amplitude of the radio wave is better known as AM or amplitude modulation. The magnitude of these variations determines the strength of the radio wave. The difference between the carrier and actual signal transmitted is called deviation and the amount from positive peak to negative peak is the bandwidth.

The amount of deviation of an audio signal determines the frequency response and dynamic range. The typical FM signal has a frequency response of about 50–15,000 Hz and a dynamic range of more than 90 dB. The deviation is a function of the amplitude of the audio signal and is usually measured in kilohertz (kHz). Typical values of deviation in wireless microphone systems range from about 12 kHz to 45 kHz, depending on the operating frequency band.

Wireless microphones operate in a total radio field consisting of direct waves, indirect waves and radio noise. Direct waves are those that travel by the shortest path from the transmitter to the receiver. The indirect waves or reflections are the result of radio waves bouncing off surrounding surfaces. This causes the wireless receiver to receive multiple signals from the wireless transmitter. This is called multipath distortion and can cause drop-outs or distortion of the audio signal.

## Radio Waves

The radio waves that arrive at a receiver by direct and indirect paths have different amplitudes related to the strength of the original source. This is due to the amount of delay and loss due to reflections, material attenuation and the total distance the transmitter is from the receiver. Radio waves, like sound waves, become weaker as they travel away from their source, at a rate governed by the inverse-square law: at twice the distance, the strength is decreased by a factor of four.

Reflective waves are usually present indoors and bounce around the environment until they are weak and nondirectional. Reflected radio waves ultimately contribute to ambient radio noise produced by many natural and man-made sources. Radio noise is across a wide range of frequencies and the strength of ambient radio noise is relatively constant in a given area and does not diminish with distance.

Radio waves are also affected by the size and composition of objects in their path. Generally, the size, location, and quantity of metal in the vicinity of radio waves will have a significant effect on their behavior. In particular, they can be reflected by metal if the size of the metal object is comparable to or greater than the wavelength of the radio wave.

Nonmetallic soft surfaces such as the human body cause significant losses to short radio waves when they pass through it. The amount of attenuation or loss is a function of the thickness and composition of the material and also a function of the radio wavelength. Generally, dense materials produce more loss than lighter materials.

## Receivers

In FM, the receiver reverses the process and converts the carrier signal into a usable audio signal (Figure 8.11). A process called heterodyning filters the carrier and then converts it to an intermediate frequency (IF). After filtering in the IF section, the signal is amplified and then sent to an FM demodulator, which separates the superimposed audio signal from the RF carrier, and converts it into a usable microphone signal again.

Stand-alone or rack-mounted receivers are the most common for use in applications where the receiver is usually in a fixed location.

**Figure 8.11** Wireless receiver units.

Electronic news gathering (ENG) and electronic field production (EFP) receivers are portable receivers that usually operate off batteries. These receivers are typically used in remote work for broadcast applications where a small lightweight receiver is either attached directly to the camera or placed on the body of the camera person like a beltpack.

Rack-mounted, card-frame style receivers are typically used in large multisystem venues where numerous systems are operating simultaneously.

*Large-scale entertainment events and parades are often choreographed and timed to prerecorded music and sound tracks. Handheld wireless microphones are used to sing over recorded tracks but also are used for props and communications. There are many situations where a live performance is not practical or possible and an entertainer or vocalist would choose not to sing live. Often a vocalist may wait till the last minute to decide and during rehearsals work with tracks and the live mix. The handheld wireless microphone is active and the sound mixer will use it when appropriate, such as at the end of a performance for the "thank you."*

*Wireless microphones are also used for off-air communications. The wireless microphone or bodypack transmitter is a direct communication path to the sound mixer, who can direct an off-air path to the director and producer. This has proven to be a critical and valuable tool during rehearsals.*

*Small wireless transmitters were a perfect solution for Larry Estrine at the Sydney Olympics. Hands-free continuous communications with the cast was essential for inexperienced actors.*

*A low-power FM transmitter was brought in so that performers would have a way to hear a sound mix and cueing system without the delay of the PA system. Large events with PAs often experience acoustical delays at different areas of a venue. To compensate for any delay in the sound, all performers were supplied a "walkman" type radio receiver so they all heard the same sound track at the same time with producer talkback.*

*A speaker podium rose from underground and there were worries that the wires could get snagged and broken, so the podium microphone was split three ways: PA, broadcast and RF transmitter.*

Computer interface software is available in most high-end receivers and provides a graphic display of all internal settings and RF status (Figures 8.12 and 8.13), and enables downloading and uploading frequency groups to and from the receiver, as well as adjustment of a variety of operating modes. Ethernet and other systems provide PC interface for the MAC and Windows® operating systems. The receiver can be also used to perform a scan to identify RFI and find clear operating spectrum.

**Figure 8.12** Transmitter status on laptop.

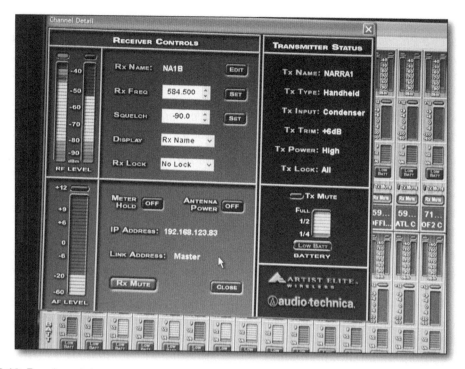

**Figure 8.13** Receiver status.

# Frequency-agile Units

In the past several years, frequency-synthesized wireless equipment, commonly called frequency-agile units, has really changed the industry. Synthesized wireless microphones are popular with field news crews, location film and TV production companies because of the ability to rapidly change to a new operating frequency. With frequency-agile systems, there are multiple channels to choose from, allowing you to change channels on-site. High-end wireless systems from Audio Technica, Shure and Sennheiser can accomodate between six and eight different frequencies in a television channel in the US (6 MHz bandwidth per TV-channel) and between eight and ten per television channel in Europe (8 MHz bandwidth).

Most frequency-agile wireless microphones use a process called phase locked loop (PLL) synthesis, which is a method of generating a stable radio carrier but which can be shifted to other adjacent frequencies within a specific range of frequencies.

Phase noise also affects channel spacing; high-performance circuits with 25-kHz channel spacing are usually considerably more difficult to implement and more expensive than those with 100 kHz or more spacing.

# Intermodulation

If two transmitters are within several feet of each other, the transmitter output stages can mix the two signals, a phenomenon called *intermodulation*. On the other end, two strong signals getting into the receiver can generate intermodulations in there, which have not been "on air" before. These receiver intermodulations are more common than the inter-transmitter intermodulations. Radio signals will combine to produce IM signals through second, third, fourth, fifth, sixth, and even seventh-order combinations.

If two signals are present at the same point in a circuit component, a sum and a difference signal will be produced. This is called second-order intermodulation or IM. This primarily occurs at the first mixer stage in the receiver and between the transmitters themselves. In a receiver, the interfering frequencies will pass right through the front-end filters in the receiver and generate the IM signal in the first mixer.

**Figure 8.14** Rack of AT5000 Receivers at Opening Ceremonies 2004.

IM can also occur from mixing three signals, or from the mixing of a signal and a second harmonic of another signal. This third-order IM can cause real problems that cannot always be prevented by highly selective receiver front-ends. In the case of third-order IM, it is possible for the interfering signals to be simultaneously close together and close to the receiver's operating frequency. When setting up wireless systems with more than two transmitters, you have to make certain that NONE of the generated 3rd-order intermodulations will be close to any of your selected frequencies. For a three-channel system, there will be six intermodulation frequencies, whereas a 20-channel system has to deal with 380 intermodulations. The set-up of a proper frequency plan is the most important factor for a successful operation.

For a number of reasons, synthesizers have far more phase noise than crystal-controlled oscillators. Although synthesizer noise is not the only cause of this problem, it is one of the most common. A poor system SNR will be clearly revealed, however, by annoying "fizzing" sounds or "noise tails" at the end of words. This effect is also sometimes called "breathing." The problem is especially pronounced when certain kinds of background sounds are present, such as waves breaking, train and subway rumbles, elevator noise and heavy breathing by the performer.

## Radio-frequency Interference

Radio-frequency interference (RFI) is generally an undesired RF signal that causes noise or distortion, limits operating range and causes drop-outs. Interference can result from external RF signal sources such as television station broadcasts, wireless microphones, intercom, IFB, remote control, communications, video signals, or digital data transmissions. Wireless mic systems utilize the same frequency bands and the same shared spectrum space as other users. Interference in a simple single-channel wireless usually results from a signal on the same carrier frequency from an external RF signal or RF noise near the receiver.

RFI problems are generated by multiple receivers and transmitters operating multiple systems in the same location. In multi-channel wireless systems, you add the 3rd-order intermodulation products to your potential interference sources. Spacing the channels very far apart—as you might hear sometimes—is NOT a solution. Successful frequency plans usually have a dense group of frequencies (minimum distance should be twice the bandwidth, so 200 kHz for a UHF-wireless system with 100 kHz) at the lower end of the used frequency band and a second group at the higher end with a big gap in the center. Most of the intermodulations will appear in this gap and outside of the used frequency band.

However, this also restricts the number of systems that are usable in any one location. If the user wants a large number of channels in one location, then some of the channels are going to be placed relatively close together. Third-order calculations can be done to select frequencies that minimize the problems of receiver interference. Most manufacturers have web sites to help select compatible frequencies in locations by zip codes. More advanced receivers can do an "autoscan" to find compatable frequencies automatically.

## Diversity Reception

Diversity receivers should be used in all broadcast applications to minimize the potential for drop-outs due to the presence of external sources of radio frequency interference at large sporting events.

The diversity circuit itself does not increase the range but rather it makes the effective range more trouble-free.

Direct radio signals and signals that have reflected off metal objects produce a condition known as multipath. Multipath is a "phase cancellation" where the direct radio signal and a reflected radio signal combine in the receiver and the two signals are out of phase, anywhere from 0 to 180°.

The reflected wave has a time difference related to the original signal, which causes it to partially or even totally cancel the original signal. The result is a noise hit or sometimes a complete loss of signal commonly called a drop-out. The null in the signal (drop-out) will happen in the physical location where the mixing occurs.

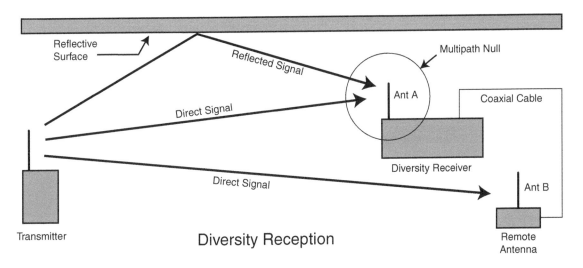

**Figure 8.15**

Simply moving just a few feet away often will cause the multipath to disappear. A diversity system overcomes the multipath problem by having two antennas in separate locations, minimizing the possibility of having a multipath cancellation at both locations. The diversity technique looks for the best signal and selects that antenna and attenuates the antenna with the poorer signal.

There are two basic forms of diversity circuitry in use today and numerous variations on these basic designs. All have advantages as well as disadvantages, depending on the quality of the receiver circuitry.

*Active-switching diversity or true diversity.* The most common form of diversity in use today is active-switching diversity, which is two antennas and two complete receiver sections along with a comparator circuit that monitors the receiver with the best signal. When one receiver begins to lose signal, the comparator switches the audio to the other receiver. (See Figure 8.16.)

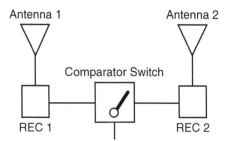

Figure 8.16 Dual receivers with antennas and a comparator circuit that determines the best signal and feeds it to the output.

Although this method is effective, the design can have problems from receivers that are not properly matched or that drift over time. These designs are expensive to produce because two receivers including critical circuitry such as RF and IF filtering must be made. When electrical components age, each receiver can drift further, compromising the performance of the system. This usually leads to noise from one of the receivers or from the switching network.

There are several variations of this basic design on the market today. Some of these include soft switching between receivers to eliminate the harsh transients that can occur when a poor signal is suddenly replaced by a good signal. Either type of design can be effective, provided the receivers are matched, and the RF and IF filtering is not compromised.

Antenna-switching diversity uses a single high-quality receiver, along with two isolated antenna inputs fed to a signal-strength comparator. When the signal begins to deteriorate in the primary antenna, the comparator selects the other antenna to try to improve the signal strength. Since this diversity technique only requires one receiver, the designer will often make a receiver with superior RF and IF filtering for the same given price point as the twin receiver diversity.

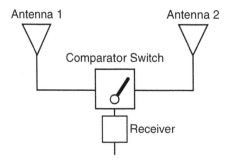

Figure 8.17 Dual antennas with a single receiver and a comparator switch that determines the best signal.

This single high-quality receiver will often provide superior performance in sensitivity, selectivity and IM rejection. However, the antenna-switching circuitry switches blindly regardless of whether or not the current signal is the strongest of the two. The result can be a switch to a weaker signal than the first, which adds noise and possible drop-outs into the signal path.

## Antennas

Place antennas properly. Antennas should be mounted away from large metal objects or surfaces, which cause reflections that can reduce signal strength. Antennas should also be kept away from sources of RF energy, such as computers, digital devices, AC power equipment, etc. For optimum results, the transmitter(s) should be in a clear line-of-site to the receiver antennas.

For best diversity performance, antennas should be placed from half-wavelength to one full wavelength apart (1 to 2 feet for UHF). Antennas are measured in terms of the wavelength of the signal that they are designed to pick up. Most UHF wireless systems use a half-wave, meaning that it is half of the length of the radio wave being received. Wider separation between antennas does not offer significant benefit, except when mounting at opposite sides of a very large stage, for example. For diversity wireless systems, both antennas must be located in the performance area, and connected to the receiver via separate coaxial cables.

*Ground plane antenna ATW-A20.* (See Figure 8.18.) Tuned elements improve antenna gain over standard ¼-wave whip antennas. The ground-plane antenna has an omnidirectional circular coverage pattern and provides about 3 dB extra RF gain. Ground-plane antennas are tuned by frequency bands, with different lengths for different RF bands.

**Figure 8.18** Ground plane antenna ATW-A20.

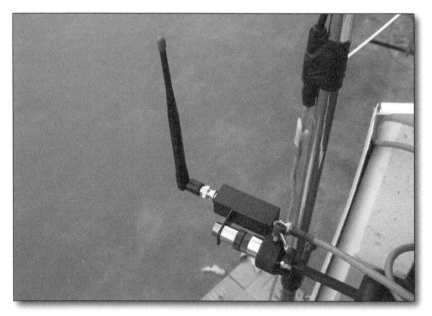

**Figure 8.19** Mono pole antennas with band filter.

**Figure 8.20** Log Periodic Dipole Array (LPDA) ATW-A49. This is a multielement, directional antenna that operates over a wide frequency range (typical bandwidth is 500 to 800 MHz) with a typical gain of 4–6 dB over a dipole antenna. The pickup pattern is similar to a cardioid microphone, which reduces pickup from rear and to a lesser degree the sides.

Sometimes called a *paddle*, a *log periodic antenna* is commonly used as a single antenna feeding a multichannel wireless system operating on a very wide bandwidth of frequencies. An LPDA antenna should be mounted so that it is not close to a nearby reflecting surface. At UHF frequencies, there is usually not a problem with placement, but at VHF frequencies there can be some limitations in indoor use. (See Figures 8.19 and 8.20.)

**Figure 8.21** Powered antennas ATW-A54P; A64P.

*Powered Antenna.* With powered antennas, an amplifier located in the antenna boosts the signal before the long cable run between antenna and receiver. This antenna requires the antenna to be powered by the wireless receiver or distribution amp to operate. (See Figure 8.21.)

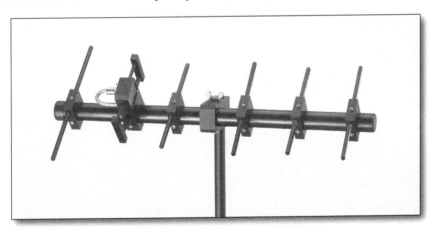

**Figure 8.22** Yagi antenna.

*Yagi beam antennas* are a multielement, highly directional antenna that operates over a limited frequency range (Figure 8.22). A Yagi antenna consists of a basic dipole element with other dipole elements placed at specific distances in front of and behind the basic dipole element. The element behind the basic dipole is called the reflector and the elements in front of the dipole are called directors. As more director elements are added in front of the dipole, the pattern becomes more directional. The Yagi has high rejection from the rear and a typical 3-element Yagi will produce about 3 or 4 dB of gain over a dipole. A 5-pole design can produce as much as 10 or more dB of gain over a dipole. This antenna must be aimed properly to be effective. The higher the gain, the more critical the placement.

*PWS helical antennas* are custom built and designed to have a very narrow bandwidth and be highly directional. This results in an antenna gain of around 14 dB but the antenna must be aimed precisely to be effective.

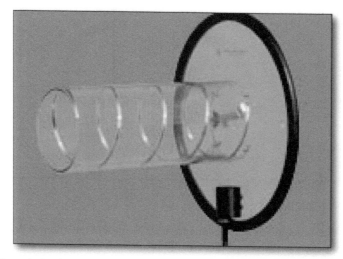

**Figure 8.23** Helically polarized antennae.

A handheld mic is continuously moving and reorienting the transmitter antenna. A helically polarized antenna picks up a constant amount of RF no matter what polarity the transmitter is. "The antennae themselves are the most important part of the whole system," said James Stoffo, president of Professional Wireless Systems (Orlando, Florida). Professional Wireless Systems provided the frequency coordination and equipment for the 2005 Superbowl in the Jacksonville Stadium. Jacksonville Stadium is right next to two huge RF towers. "When you're that close to an FM transmitter, you're going to get spurs, harmonics, and intermods. For the Super Bowl halftime show we used helically polarized antennae, because they just don't drop out."

## Antenna Cables

For proper performance the impedance of the cable connecting the antenna to the input on the receiver should match, typically 50-ohm RG-58 cable for shorter distances (below 20 feet) and RG-8 for longer runs. An impedance mismatch causes some of the signal to be reflected back into the cable, resulting in a reduction of signal level. A common mistake on a television shoot is to use RG-59 which is video cable (75-ohm).

You have to be careful when using more than two amplifiers in a line, as the distribution amplifier must be able to provide enough current on one hand and the noise floor might come up on the other hand. So the correct placing of the amplifiers becomes very important.

Different types of cable are rated according to the amount of signal loss over a specified length. These ratings are specific to the frequency range of the signal being passed (VHF or UHF), particularly on long runs between the antenna and the splitter. RG-8 is a very low-loss cable with loss of 2 dB per hundred feet, at 700 MHz.

Minimize the number of connection points by using proper lengths of cable. Each connection between two sections of cable results in some additional signal loss. It is not necessary for both antennas on a diversity receiver to be connected to identical lengths of cable. Ideally, each antenna should be connected to the minimum length of cable necessary to reach the receiver.

The maximum length of cable is dictated by how much the antenna amplifier can boost the signal. If antenna boosters are used, the cable length can usually be longer, depending on how much gain the booster provides. If antenna amplifiers are being used, mount the antenna directly on the input of the first amplifier and use one length of cable to go from the amplifier to the second antenna amplifier directly to the receiver. Up to two amplifiers can be used in a design, one located at the antenna, and one in the middle of the cable run, permitting total cable runs of up to 500 feet when the appropriate type of coaxial cable is used. You have to be careful when using more than two amplifiers in a line, as the distribution amplifier must be able to provide enough current on one hand and the noise floor might come up on the other hand. So the correct placing of the amplifiers becomes very important.

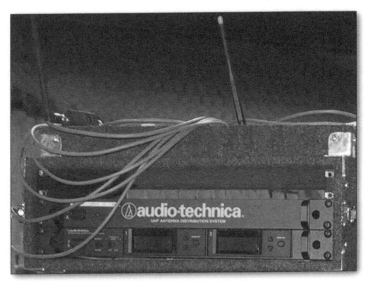

**Figure 8.23a** Antenna Distribution AEW-D550C; D660D. Typically tuned to specific RF frequency band, this active device provides unity gain. Use single set of antennas with multiple receivers.

Active device systems are typically tuned to a specific RF frequency band that provides unity gain. This allows the use of a single set of antennas to feed multiple receivers.

If a cable run longer than 500 feet is called for, it is better to locate the entire receiver closer to the performance area and run a long audio cable to the mixing console. A long audio cable (balanced and line level) is much less likely to pick up interference than a long antenna cable.

# Wireless Communications

Wireless intercom systems have become standard in all areas of broadcasting. For entertainment shows the stage manager plays a major role in the flow of a production. A stage manager will require a duplex system because it provides uninterrupted hands-free communications but requires two continuous carriers to transmit and receive. Additionally, many PL systems will have a switch so the wireless stage manager can use the PA for announcements and directions.

Wireless PL channels are needed for RF cameras. RF cameras are used extensively in sports and entertainment programs and require a continuous listen of the director calls. At the director's communications panel, there will be two channels for camera communications. It is desirable to keep the hardwired stationary cameras on one channel of communication and the RF cameras on another.

The master station connects to a channel of the PL system either with a four-wire interface or matrix ports systems such as the RTS Adam or Reidel. The beltpack provides dual audio monitoring, with two volume control knobs for independent intercom and program level adjust, or for separate level control on intercom channels one and two.

Frequency-agile UHF wireless intercom system with operating frequency bands between 470–740 MHz (or up to 862 MHz in Europe).

In-ear monitors. When the UHF band became available, manufacturers really started promoting in-ear monitor systems. Roving reporters use a wireless handheld mic or a bodypack microphone with a wireless IFB. The problem is that the IFB in-ear monitor receiver is in close proximity to the wireless microphone transmitter. Large award shows have used up to 75 wireless microphones plus an additional 20 channels of RF PL for stage managers, cameras and audio intercom and up to 12 in-ear monitors and IFBs. The key to the successful integration of so much RF is a proper frequency co-ordination as discussed in the following section.

# Frequency Coordination

Government policy dictates which frequency ranges can be used for specific uses. Wireless frequencies are shared with TV stations, communications equipment and a large number of ENG wireless microphone systems. Outside broadcasts require frequencies for microwave links, talkback, wireless cameras and handheld radio. Government regulations also set strict technical requirements for wireless on maximum transmitter power.

Because of the large number of users, there is a significant amount of frequency sharing, so there is a chance that someone else in the area might be using the same frequency as your wireless system. There's an FCC Web page, which allows you to go view licensed frequencies and television stations and radio stations in any given area, but it does not list theaters, churches or other heavy users of frequencies. With the new frequency-agile equipment, you've got options at least.

The law is that you cannot transmit within an active television carrier. Every major metropolitan area has an SBE (Society of Broadcast Engineers) frequency coordinator to make sure that when someone comes in—a news crew or some other wireless operators—they know what the active stations are and not to transmit on any of those frequencies. James Stoffo is the frequency coordinator for central Florida, and when there's a major event like Daytona 500 or a big launch at Cape Canaveral or when CBS or ABC or NBC comes to Orlando, Daytona or Ocala I get phone calls and spend a couple of hours coordinating frequencies to make sure that no one's stepping on or causing interference with anybody else. This isn't anything new for the guys in television or radio!

A major challenge over the next three years is the fact that the RF band where wireless mics currently operate will become more congested: UHF band 470 to 806 MHz. The FCC has auctioned off over 100 MHz of spectrum and the users have lost close to 150 MHz, from 470 to 512 and 698 to 806 MHz. The whole inventory is now squeezed into from 512 to 698 MHz, whereas before there was an extra hundred megahertz worth of frequency allocation.

As DTV stations come on the air and analog telecasts remain active, the available spectrum will decrease dramatically. You will have parallel programming carriers, digital and analog TV, which leaves very little room for wireless ENG crews and anybody who has a current UHF band 470-to-806 MHz wireless system.

## Spectrum Allocation in North America

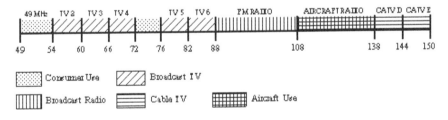

**Figure 8.24** Band 49 MHz–54 MHz Common band for such things as toys (walkie talkies), cordless telephones. Radio frequency interference (RFI) is produced naturally by the sun during the eleven year sun spot cycle.
Band 54–72 MHz TV channels 2–4.
Band 66–72 MHz channel 4.
Band 72–76 MHz hearing assistance systems.
Band 76–82 MHz channel 5.
Band 88–108 MHz commercial FM radio broadcasting.

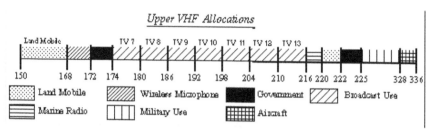

**Figure 8.25**  Band 150–168 MHz Police/Fire, and business.
Band 169–172 MHz Traveling band used for wireless microphones.
Band (174–216 MHz) VHF broadcast upper television broadcast band channels 7–13 and
wireless microphones.

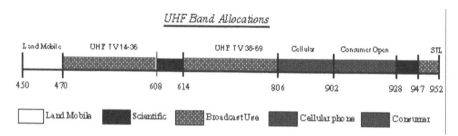

**Figure 8.26**  UHF Low (450–469 ) commercial users, business communication.
UHF Mid (470–806 MHz) television broadcast band. Wireless microphones
UHF (746–806 MHz) TV channels 60 to 69 are being re-allocated
UHF High (902–928 MHz) Operation is license free. Like the 49-MHz band, 902–928 band
allows for operation of transmitters up to 50 W of output power.
UHF High (947–952 MHz) STL studio transmitter link for high-power data transmitters
for broadcast use.

## Digital Radio Microphones—Latency Issues

Most broadcast technologies have gone digital over the last decade, and have benefited as a result, but
digital wireless microphones and cameras have delay problems that make them difficult to use. Radio
microphones need real-time transmission and compressing/companding  converting the signal from
analog to digital and back again causes delay. With digital wireless this problem is most noticeable
with lip synchronization in live performances and when you mix standard wired microphones with
digital wireless.

## Batteries

Weak or worn-out transmitter batteries are a common cause of wireless problems, including complete
failure, poor range, distorted audio and interference. High-quality alkaline batteries will provide 8 to
16 hours of transmitter operation. Most other types of batteries will have much shorter life, and some
may cause other problems.

# Waterproofing RF connectors

There are different ways to protect and waterproof your connections. Wrap the entire connection once or twice with 3M Scotch Super 88 electrical tape. Be sure to allow a significant overlap of each turn. It's also to best to wrap the tape up the jacket and the opposite way of intended water flow. This is done to prevent the electrical tape from wicking water in. On the last turn, fold the edge of the tape over and press it down to prevent it from unwrapping. Make sure there are no air cavities or openings in the tape to permit water to penetrate. The tape also will help prevent the connection from coming loose due to vibration.

# Active Switching Diversity—Systems Review

## *The Audio Technica AEW-R5000 Wireless System*

This system is a dual-receiver design with two independent UHF receivers with true diversity reception and automatic switching in a single housing. Dual compander circuitry processes high and low frequencies separately for higher audio quality. Computer interface with Ethernet 10BaseT on RJ-45 connector for monitoring and controlling system parameters has become standard. PC compatible, proprietary software IntelliScan™ automatically finds and sets best available frequencies on all linked receivers. The white-on-blue LCD information display is high visibility. Each component can store up to five preset configurations, with customized "names." It has front-mount antenna cables and connectors for two flexible UHF half-wave antennas.

### Specifications and Features

Advanced digital Tone Lock™ squelch effectively blocks stray RF; the digitally encoded tone also communicates transmitter data for receiver display.

Link cable: Link and coordinate multiple receiver channels. External mute available on 1/4" (6.3 mm) TS unbalanced phone jack allows user to mute the system quickly and easily.

Image Rejection: 60 dB typical signal-to-noise ratio: 115 dB at 40 kHz deviation (IEC-weighted), maximum modulation 75 kHz Total Harmonic Distortion: <=1% (10 kHz deviation at 1 kHz); sensitivity: 20 dBµV (S/N 70 dB at 5 kHz deviation, IEC-weighted); Intermediate Frequency: 65.75 MHz, 10.7 MHz.

Audio output is microphone level, 25 mV (at 1 kHz, ±5 kHz deviation, 10k ohm load)

Instrument: 50 mV (at 1 kHz, ±5 kHz deviation, 10k ohm load) Outputs are transformer isolated and balanced Audio Output Attenuator (ATTN) : Three-position switch: 0 / –6 / –12 dB

Output Connectors Microphone: XLRM-type Instrument: 1/4" (6.3 mm) TRS balanced phone jack

Headphone Output Connector: 1/4" (6.3 mm) TRS ("stereo") phone jack

The AEW-T1000 UniPak™ Transmitter is enclosed in a rugged, ergonomic metal body. Programmable features on soft-touch controls include control of switchable RF power 10 mW/35 mW power output.

(Maximum output: 220 mW + 220 mW into 32 ohms) The dynamic range microphone input is >=110 dB, A-weighted, Instrument: >=100 dB, A-weighted. Net Weight (without batteries) 4.4 oz (125 g).

## Sennheiser EM550G2 Wireless Twin Receiver

The EM 550 G2 is a true diversity twin receiver with HDX compander system with > 115 dB(A) signal-to-noise ratio. Pilot tone squelch for eliminating RF interference when transmitter is turned off.

Four equalizer presets, transformer-balanced audio outputs. Nine frequency banks with up to 20 directly accessible presets each. Automatic frequency scan feature searches for available frequencies in all frequency banks.

Integrated antenna splitter for cascading several EM 550 G2 without the need for an additional splitter.

Antenna boosters are powered via the antenna cables. Menu operation via two-color backlit graphics display (green = OK; red = warning) 4-step battery status display of transmitters; Mute indicator for transmitter.

Soundcheck mode for checking AF and RF conditions. Adjustable headphone monitoring output.

Lock function avoids accidental changing of settings.

## Research

Fox, David, *Wireless Microphones*, Audio Technica, Sennheiser, Shure, Lectrosonics.

Stoffo, James, *Professional Wireless Systems* (Orlando, Florida).

# Index

**A**

A-B, or switch, 40
ABC Sports, 54-55, 188-189
AC (alternating current), 58, 64, 116, 141, 145-146, 207, 227
Academy Awards, 10-11, 141-142
Active Switching Diversity, 235
    Active-switching, 225
ADAM, 129, 131-132, 137, 231-232
    ADAM Intercom, 131
Adams, Peter, 184
AES, 149-150, 153-154, 158-159, 209, ix
    AES10, 60
AIC (audio in charge), 11
Aldous, Fred, 7
AMEK, 53
American Broadcasters, 1
American Idol, 9
Analog Audio, 56-57, 59, 61, 64, 86, 131, 141, 143-146, 153-155, 206
    Signals, 141, 143-145

Analog Volume Unit, 61
Analog-to-digital-to-analog, 151-152, 156, 158-160
Announcers, 5-6, 10-13, 20, 27, 55, 81, 83, 93, 95-96, 99-100, 105, 107-108, 110, 125, 136, 146, 175, 183, 201
Antenna, 9-10, 211-212, 218, 225-232, 235-236
    Cables, 230, 235-236
    Distribution AEW-D550C, 231
    Switching, 226-227
Aphex Compellers, 92
Aphex Electronics, 153-154, 161
Arrays, 198
Assignable Control Functions, 50, 73
AT microphone, 180
Athens Greece, 56
Athens Olympics, 40-41
Atlanta, 4-5, 120-121
    Atlanta Braves, 4
    Atlanta Motor Speedway, 5
Atmosphere, 5, 20, 25-29, 31, 34-37, 46, 55,

83, 91-92, 95-96, 110-111, 146, 165-166,
182-186, 190, 192-193, 208
    Microphones, 91-92, 95-96, 110-111, 184
ATTN, 235
ATW-A20, 227
ATW-A49, 228
ATW-A54P, 229
Audible, 30, 40, 76, 85, 92, 168, 190
Audience, 2-3, 5-11, 20, 28-29, 34-38, 42, 47,
    77-78, 95, 164, 182, 184-186, 199, xi
Audio,
    Assistant, 13, 16, 19, 22, 24, 99-102,
        104-105, 107-111, 113-118, vii
    Crew, 27, 59, 101
    EIC, 15
    Mixer, 2-3, 10-11, 15, 19-21, 24, 27, 34-35,
        49, 76-78, 92, 99, 110-111, 114-115,
        135-136, 186, xi-xii
    Output Attenuator, 235
    Positions, 10-11, 16, 110-111
    Set-up, 101
    Technica AEW-R5000 Wireless System, 235
    Technica AT825, 193
    Technica AT849, 25
    Troubleshooting, 113, 116
Audio-follow-video, 78
Audio-Technica, 181, 189, 209
Australia, 164
Auxiliary, 57, 69, 73-75, 79, 81, 87, 90, 134
    Auxiliary Sends, 57, 69, 73-75, 79, 87, 90
    Effect Sends Entertainment, 74
A-weighted, 235-236
AZedit, 132


**B**

Balanced, 11-12, 14, 77, 85, 128-129, 135, 142-
    145, 181-182, 207-208, 231-232, 235
    Audio XLR, 208
    Output, 181-182, 208, 235
Balancing, 10-11, 16-17, 20-21, 29, 39-40
Ballistic, 64-65
Band, 30, 42, 47, 69, 73, 89, 139-140, 163, 218-

219, 224, 228, 231-234
Bandpass Filter, 69
Bandwidth, 5-6, 66, 71, 131-132, 138-139, 150-
    151, 154-156, 158-161, 218-219, 222, 224,
    228-230
BASE-T, 131
BASE-TX Ethernet, 131
Basics of,
    Microphones, 166
    Mixing Consoles, 56
Basketball, 10-11, 20, 25-26, 99, 102, 111, 113,
    175, 185, 203, 207
Batteries, 179, 212, 214, 220, 234-236
    Weak, 234
Beatles, 1
Bell Labs, 149-150
Blumlein, Alan, 191
Bodypack, 213-215, 220-221, 231-232
Boom Operator, 1, 13-15, 195
    Hollywood, 14
Booth, 16, 100, 102-110, 118, 130-133, 137,
    161
Boy Scout, 117-118
BP, 130-131, 134-135
    BP-325, 133
Bravo, 4
Broadcast, 1-2, 4, 8-10, 15-17, 19, 27-28, 30,
    39-40, 42, 45-46, 51-52, 59-61, 75, 80, 83-
    85, 95-96, 99-100, 108, 114-115, 119, 125,
    130-133, 137, 141-142, 145-146, 151-154,
    161, 177-178, 183, 211, 220-221, 224, 232-
    234
    Systems, 151-154
Broadcasters, 1, 4-6, 10, 15-17, 38-42, 46, 51-
    52, 54, 93, 113, 145-147, 153-155, 160, xi
Broadcasting, 5-6, 32, 51-53, 74-75, 85, 92,
    108-110, 113-114, 164, 192, 211, 231-233
Buck Owens Variety Show, 30
Butler, Mark, 22-23


**C**

Cable—Hooking It All Up, 141

Cables, 13, 16-17, 21, 104-105, 114-116, 130-131, 141-146, 151-152, 156-157, 175-176, 207-208, 227, 230, 235-236

Cabling, 9-10, 16, 102-103, 120-121, 137, 141, 145-146, 154-155, 188-189, 207

Calrec, 51-52, 76, 93

Cameras, 2, 5-8, 14-17, 19, 24, 26-27, 35, 49, 59, 65-66, 78, 91, 95-96, 108, 112-116, 119-122, 125-126, 130-132, 163-164, 172-174, 183-186, 188-192, 194-195, 197, 207, 214, 220, 231-232

Camera C14, 26

Camera C7, 26

Camera C8, 26

Camera C9 SSM, 26

Cape Canaveral, 232-233

Capital Cities, 4

Capturing, 7-8, 24, 28, 30, 35-36, 163-166, 175-176, 182, 191, 198, 208, 214, xi

Car, 5-6, 114-115, 174, 183-185, 203, 209, 211-212

Carbon, 134

Cardioid, 5, 169-171, 174-175, 177-178, 186-189, 191-192, 194, 228

CAT, 141, 161

CAT5, 126, 145-146

CAT-5, 154-155, 158-159

Caution, 35-36

CBS Mobile Unit, 5-6

CBS OB, 156

CBS OBV, 54

CCTV, 141

CCU, 207

CDP, 138

Channel Insert, 80-82

Chicago, 4, 149-150

Cigarette, 96-97

Circle, 5-6, 32-33, 40-41

Circle Surround, 5-6, 32-33, 40-41

Cleaning, 69, 96-97, 157-158

Optical Fibers, 158

Clear-Com, 120, 125, 153-154, 158-159

Close-up, 25, 71, 91-92, 149, 188

CNBC, 4

CobraNet, 154-156

Codec, 131-132

Color, 20, 22, 34-35, 84, 102, 108, 168, 177-178, 208

Commercials, 31, 39, 42

Communication Systems, 106, 119-121, 158-159, vii, xi-xii

Communications, 2-3, 8-9, 14-16, 24, 27, 60, 99-100, 102-105, 108-110, 114-115, 119-133, 135-136, 141-142, 158-159, 161, 220-221, 224, 231-232, xi-xii

Intercoms, 14

Using Voice-Over-Internet Protocol, 130-131

Compression, 30, 42, 47, 50, 72-73, 81, 92-93, 125, 216-217

Controls, 72-73

Computer, 65-66, 126, 137, 139-140, 158-159, 221, 235

Condenser, 175-176, 178-182, 186, 207, 214-215

Connection, 30, 60, 93, 102-103, 117, 124, 126, 131-132, 143-147, 150-151, 155, 175-176, 181-182, 207, 230-231, 234-235

Connectivity, 114-115, 141, 145-148, xi-xii

Connectors, 51-52, 102, 114-118, 129, 143-146, 151-152, 156-158, 190, 207, 234-235

Console Inputs, 57, 179

Consoles, 2-3, 7-8, 15, 21, 29-30, 49-58, 60-61, 63-66, 68, 71-73, 75-85, 87-92, 96-97, 102-103, 105, 141-142, 153, xi-xii

Contact Microphone, 203-204

Contrast, 99

Control, 3-6, 12-15, 21, 27-29, 39, 49-52, 55-59, 65-66, 70-78, 80, 82-92, 94-95, 105-108, 113, 123-128, 130-131, 133-134, 136-138, 149-150, 153-155, 158-159, 161, 188-189, 198, 207, 216, 224, 231-232, 235

Conversions, 61-62, 91, 151-152, 158-160

Copper, 44, 59, 99, 102, 124, 129, 141-154, 156,

190, 207

Copper-based Ethernet, 154-155

Core Elements, 27-28

Core Elements of a Sound Mix Viewers, 28

Corning, 149-150

CPU, 88

Crackling, 114-115

Creative, 2-3, 10-11, 23, 27-29, 31, 198, 208

Creativity, 5, 23, 165, xi-xii

Crew Communication, 23

Crowd, 10-11, 26, 34-37, 46, 102, 105, 175-176, 203

Microphone, 102

Crown PZM, 188-189

C-size, 212

**D**

D660D, 231

DA, 81

Danish, 182

Davidson, Joost, 44

Davies, Tim, 47

DBX, 92-93

DC, 114-115, 120-121, 124, 136-137, 181-182

D-Cams, 44

D-D, 156

DE-9, 129

Decca Recording Studio, 197-198

Decca Tree, 197-198

Decorrelation, 35-36

Designing, 51-52, 80, 88, 186

Designs, 9, 49, 73-74, 76, 86, 88, 108, 120, 138, 158-160, 172-173, 192-193, 208, 225-226

Digging, 203

Digi Cart, 93-94

Digital,

Equipment, 60-61, 91, 145, 158-160

Mixing Consoles, 15, 21, 49, 55-56, 85, 90-92

Mults, 59, 141-142, 145-146, 153-154

Mults Copper, 153

Radio Microphones, 234

Systems, 124, 126-128, 130-131, 141, 151-155, 158-160

Direct Reflections, 207

Direct Sound, 166, 174, 188-189, 207

Directional, 158-159, 167-176, 186, 188-189, 191-192, 194, 205, 228, 230

Microphones, 167, 169, 174-176, 191-192

Director, 1, 10-11, 15-16, 19-21, 23-24, 27-28, 49, 85, 100, 109, 119-122, 132, 138-140, 184-185, 220-221, 230-232, xi-xii

Bill McCoy, 122

Display, 50, 59, 65, 68-69, 73, 87-89, 92, 221, 235-236

Distance Factor, 171, 174

Distortion, 62-66, 76, 114-116, 150-151, 175-176, 182, 211, 216-219, 224, 235

Diversity, 224-227, 230-231, 235-236

Diversity Reception, 224, 235

Dixon, Bob, 15-16, 33, 38-39, 164

Dodson, Doug, 51-52

Dolby, 5-6, 31-32, 35-36, 38, 40-42, 44, 46, 94-95

Digital, 31

Pro Logic II, 40-41

DP, 42

E, 94-95

Labs, 35, 40-41

PLII, 44

Pro Logic II®, 32, 38, 44, 46

DPA, 182

DSP Processors, 89-90

DSPs, 55, 90

DT-12, 145-146

DTV, 31, 232-233

Dual, 49, 51-52, 54, 56-57, 75-77, 154-155, 172, 226, 231-232, 235

Dual-ear, 135-136

Dutch Broadcasting Organization, 192

DVD, 31

Dynamics, 10-11, 27-30, 42, 54, 57, 65-66, 69, 71-73, 76, 86-90, 93-94, 174, 180-181

**E**

Ear-training, 10-11
Easy-to-read, 65, 135
EBU, 153-154, 158-159
Edit IFB, 137
Effects, 5, 11-14, 20, 25-31, 34, 40-41, 45-47,
    74-76, 80, 83, 85, 93-96, 105, 110-112, 139-
    140, 146, 163, 174, 182, 188-189
    Effect Sends, 74
    Effects Mixer, 11-12, 45-46
EFP, 14-15, 220
EIC, 15
Electrical Properties of Microphones, 178
    Microphones, 178
Electrodynamic, 178
Electromagnetic, 59, 95, 141, 151, 216-217
EM, 235-236
Embedders, 158-159
EMI, 151
    EMI/RFI, 151
Emmys, 10-11
Emmons, Mark, 145
Emphasizing Specific Frequencies, 30
ENG, 14-15, 213, 220, 232-233
    ENG/EFP, 14-15
Engineers, 9, 23, 51-52, 60, 165-166, 178, 197-
    198, 216, 232-233
Entertainment, 2-5, 8-10, 14, 20, 24, 28-29, 31,
    45, 55, 74, 95, 99, 119, 138, 141-142, 161,
    165, 185, 220, 231-232
Equalization, 11, 29-30, 40, 50-52, 54, 65-68,
    71-73, 86, 88
Equalizer Tip, 71-72
Equipment, 1-3, 5, 7-11, 14-17, 21, 24-25, 27,
    29, 32, 51-53, 60-61, 63-64, 85, 91-92,
    96-97, 99, 101-102, 104-105, 109-111,
    113-119, 125, 130-131, 136-137, 141-145,
    151-152, 155, 157-160, 163, 166, 213, 216-
    217, 222, 227, 230, 232-233
ESPN,
    Extreme Games, 119
    Networks, 164

Summer, 145
X Games, 130-131
Estrine, Larry, 138, 220-221
ESU, 99
Ethernet, 131, 145-146, 153-156, 158-159, 221,
    235
    EtherSound, 156
Euphonix, 71, 78
Events, 2, 4, 10-12, 14-16, 24, 30, 34-35, 47, 95-
    96, 100, 104, 110, 113, 119, 125, 133, 179,
    211, 220-221, 224
    Event-specific, 91-92, 183

**F**

Facilities Check, 27, 104-105, 116
Faders, 2, 16-17, 20, 49, 51, 54-57, 75-79, 81,
    86, 88, 91-92, 96-97
Fatigue, 11-12, 42, 142
FC Connector Styles, 157
FCC, 232-233
    FCC Web, 232-233
Fiber Connector, 150, 157
    Preloaded Epoxy, 157
Fiber Optics, 9, 16-17, 83, 103-105, 116, 129,
    141, 149-152, 158-161
    Corning, 149-150
Film, 1-3, 9-10, 12, 14, 157-158, 163-164, 172,
    180, 195, 222, xi
Filters, 38, 69, 71-72, 89, 181-182, 218-219, 223
Finney, Dennis, 5
Fire, 234
Five Aphex, 37, 95
FM, 211-212, 216-221, 230, 233
Foam, 116, 199-200
Four Wire, 124-125
FOX, 1, 5-7, 15-16, 27, 31, 33-34, 42, 112, 161,
    164, 184, 209, 236
    NASCAR, 34, 161, 209
    Network Director of Sports, 1
    TV, 5
French National Broadcasting Organization, 192
Frequency, 40, 47, 65-68, 71-72, 89, 93, 138-

140, 151-152, 166, 168-170, 178, 180-181, 188-190, 199, 208, 216-219, 221-222, 224, 227-228, 230-233, 235-236
    Coordination, 139-140, 230, 232
    Frequency-agile, 139-140, 222, 231-233
    Frequency-agile UHF, 231-232
    Frequency-agile Units, 222
Functions, 10, 21, 49-52, 55-57, 59, 64-65, 67, 71-74, 76, 84, 86-90, 92-93, 99, 107-108, 116, 127-128, 132, 134-135, 138, 145, 153-154, 158-160

**G**

G2, 235-236
Gain, 30, 54, 57-58, 60, 64-68, 71, 76, 85, 89, 92-93, 95, 153-154, 187-189, 195, 216-217, 227-228, 230-232
    Gain Structure, 58, 64-66, 76
Games, 31, 47, 119, 130-131, 145, 164, 175-177, 199, xi
General Electric, 4
Gigahertz, 218
Glaser, Greg, 147
Golf, 5, 11-12, 27, 45-46, 74, 99, 105, 114-115, 156-157, 184
Good Audio Practices, 143-144
Good Design, 132-133, 186-187
Government, 232
GPI, 78, 135, 153-154, 158-159
Grammy Awards, 9-11, 37-38, 165, 199
Grease, 96-97
Greene, Ed, 165
Ground, 10-11, 102, 116-117, 142-146, 151-152, 181-182, 207-208, 227
    Ground-plane, 227
Gymnastics Federation, 175-176

**H**

Hair-thin, 149-150
Hand Mic, 26
    Hand Microphone, 102, 105, 107-108, 179, 192-193

Handheld Microphones, 20, 25, 35, 165-166, 195, 200-201, 220
Handhelds, 214-215
Handling Cables, 156-157
HD, 31, 46, 91
HDTV, 46, 158-159
HDX, 235-236
Headphone Output Connector, 235
Headroom, 65-66, 161, 175-176
Headsets, 2, 7-8, 20, 104-105, 120-121, 124, 135-136, 163, 183
High-impedance, 61-62
High-pass Filter, 69
High-voltage, 114-115, 143-144, 207
Hill, David, 1
HMF, 68
Hollywood, 9, 14
Holophone, 198-199
Holy Grail, 190
Hooking It All Up, 141
Host Broadcast,
    Broadcaster, 10, 16-17, 38-39, 46
    Training, 16-17
Housing, 8, 134, 157-158, 166, 170, 175-176, 178, 186, 188-189, 192-194, 198, 235
Hypercardioid, 165, 169, 171-172, 181, 194-196

**I**

I/O, 141-142, 145, 153-155
ID, 126, 131, 135-136
IIEC-weighted, 235
IFB Stations, 133, 137-138
    IFB's, 160
    IFB-828, 137
Image Rejection, 235
Indianapolis, 5-6, 188-189
In-ear, 84, 138, 231-232
Innovation, 7, 209
Input, 14-16, 20-22, 27, 40, 51, 53-54, 57-62, 64-66, 73, 75-77, 80-81, 83, 89, 92, 95, 101-102, 105, 108, 114-115, 128-129, 134, 136-140, 142, 145, 149-150, 153-154, 158-160,

181-182, 195, 207-208, 216-217, 230-231, 235-236
Input A Left, 136-137
Input A Left/Right, 136-137
Input B Left, 136-137
Input B Left/Right, 136-137
Input C Left, 136-137
Input C Left/Right, 136-137
Installation, 16-17, 100, 109, 113, 119, 123-124, 129, 139-140, 145-152, 157-158, 207
Instrument, 214, 235-236
Insulation, 116
Integrating Analog, 151-152, 158-160
Intelligent Trunking, 130, 132
IntelliScan, 235
Intercom, 14, 23-24, 59, 108, 120-121, 123-128, 130-140, 153-154, 158-159, 224, 231-232
Interconnect, 9-10, 15, 60, 101-102, 133, 141-142, 145, 147, 151-152, 154
Interconnectivity, 60, 141-142, 158-160
Interference, 14, 59, 95, 103, 105, 114-115, 133, 139-141, 143-144, 151-152, 158-159, 161, 172-173, 189-190, 207-208, 216-217, 224, 231-236
Intermediate Frequency, 218-219, 235
Intermodulation, 222-224
International Gymnastics Federation, 175-176
Interruptible Feedback, 136
IP, 130-132
ISDN, 127
ISO-channel, 133
Isolation, 116-117, 145-146, 158-159
Isolation/Insulation, 116
Issues, 5-6, 14, 30, 42, 44, 116, 119-121, 147-148, 156, 172, 207, 234

**J**
Jacksonville Stadium, 143, 230
Japan, 31, 39, 44
Japanese, 37-39, 42, 44
Jaws, 1
Joost, Davidson, 44

Juggling, 30

**K**
Kevlar, 145-146
Kilohertz, 218-219
Kimwipes, 158
Klotz Digital, 158-159

**L**
Landsburg, Klaus, 37-38, 165
Large-diaphragm Microphones, 29, 177, 199
Latency, 44, 150-151, 153-156, 158-159, 234
Issues, 44, 234
Lawo, 58, 70, 79, ix
MC90, 58
Layers, 28-29, 35-36, 91-92, 150, 158, 183
Laying, 20, 114-115
LCD, 235
LCP-102, 138
Leatherman, 114
Lectrosonics, 236
LED, 4, 7, 21, 59, 78, 89, 138
Left Total, 32, 44
LFE (low-frequency effects), 33, 39, 199
Light, 21, 103, 116, 134, 141, 149-151, 153, 157-159
Lighting, 14, 24, 114-115, 120, 125, 130-132, 142-143, 151-152, 207
Lightwinder, 158-159
Lightwinder-Natrix, 153-154
Line-level, 53-54, 57, 59, 75, 105, 143-145, 158-161
Live PA, 9, 47
LMF, 68
Localization, 33-35, 191-192
Lock, 134-135, 235-236
Log Periodic Dipole Array, 218, 228
Logarithmic, 63
London, 51-52, 197-198
Long, Ed, 188-189
Long PPM, 64
Long Stereo Shotgun, 26

Los Angeles, 7-8, 51-52, 54, 145-147
    Olympics, 145-146
Low-frequency, 25-26, 30-31, 33, 38, 69, 80,
    116, 175-176, 180
Low-level, 57-59, 143-144
Low-pass Filter, 69, 151-152
LPDA, 228-229
L-R, 49
Lt / Rt – Left Total, 32

**M**

M/S, 191, 195
MAC, 131, 221
    MAC ID, 131
MADI, 37, 60, 141-142
Magneto-optical, 94
Matrix, 22, 32, 35-36, 40-41, 46, 51, 55, 60,
    79, 82, 108-109, 124-127, 130-133, 135,
    137-139, 147, 153, 158-159, 194-195, 199,
    231-232
    Audio Distribution, 138-139
    Intercom Systems, 125-126, 130-131, 138
McCurdy Radio of Canada, 126
Measuring, 60, 63-64
Metal, 116-117, 141-142, 175-176, 207, 218-
    219, 224-225, 227, 235
Metering, 16-17, 30, 35, 40, 46, 62-66, 73, 92,
    95
    Meters, 61-65, 135, 145-146, 150-151
Mic Type, 26, 48
Microphone(s),
    Applications, 182, 205
    Characteristics, 177-178
    Microphone Models,
        AT3000, 165
        AT4050, 36, 112, 177, 199
        AT804L, 26
        AT815ST, 26
        AT825, 7, 192-193, 197
        AT830R, 26
        AT835ST, 26
        AT895, 26, 183, 205-206

        AT899, 26
        M1, 26, 48
        M10, 26
        M11, 26
        M12, 26
        M13, 26
        M14, 26
        M15, 26
        M16, 26
        M2, 26
        M3, 25-26
        M4, 26
        M5, 26
        M6, 25-26
        M7, 26
        M8, 26
        M9, 26
        MC14, 26
        MC7, 25-26
        MC8, 26
        MC9, 25-26
    Phasing, 189
    Placement, 5-6, 11, 23-24, 28, 31, 35-36,
        99-100, 111-112, 165-166, 174-176,
        183, 196, 198-199, 208-209, 214
    Plan, 15-16, 24-25, 113, 186, 188-189|
    Protection, 199
    Response Pattern, 168-169
    Signal, 57, 102, 105, 110, 114-115, 191,
        195, 207, 218-219
Microwave, 9-10, 127, 149-150, 184, 211, 232
Mid-side Stereo, 195
Mix, 2-3, 5-6, 8-13, 20-47, 49-50, 53, 56-58, 65-
    66, 69, 71-72, 74-79, 81-85, 87, 91-92, 95-
    96, 102, 105, 110, 125-127, 136, 138-139,
    143-144, 165-166, 175, 182-186, 188-191,
    198-199, 208, 220-223, 234, xi-xii
    Mix Every, 40-41, 46
Mixing,
    Console, 3, 15, 20-21, 27-28, 33, 35, 43-44,
        49-57, 59, 65-66, 68, 71, 73, 76-79, 81,
        85, 90, 95-97, 102, 104, 141-142, 145,

158-159, 179, 188-189, 231-232, xi
Console Outputs, 81
Consoles Analog, 56, 61, 82
Desk, 2, 5-6, 10-11, 16-17, 20-23, 27-29, 42, 46, 49-50, 54-56, 59-60, 64-66, 71-72, 76, 80-81, 84-85, 90, 92-93, 95-96, 110, 114-115, 145, 163, vii
Surround Sound, 33, 81
MME, 51
Model IFB-325, 137
Modularity, 45, 50, 54
Monaural, 134
Monitor, 9, 14-15, 39-41, 53-54, 57, 63-64, 71, 75, 82-85, 89, 105, 130-131, 231-232
Monitor Selection, 39-40, 82-83
Monitoring, 5-6, 8, 22, 28-30, 33, 39-41, 46, 49, 53-54, 65-66, 78, 82-85, 88-89, 92, 95, 135-136, 139-140, 153-154, 231-232, 235-236
Mono-to-stereo, 29, 40-41
Morrow, Jay, 5
Most North American OB, 145-146
Most UHF, 227
Motor, 5-6, 10-11, 14, 184, 188-189, 211
Motorsports, 5, 11-12, 65-66, 99
Mounting, 116-117, 164, 187, 199-200, 227
Microphones, 199
Movie Sound, 209
Movies, 30
MS, 35, 57, 156, 191, 194-196
MU8a, 7
Mult, 101-102, 142-146
Multimode, 150-151, 158-159
Multipath, 218-219, 224-225
Multiple Mix Positions, 45
Multivenue, 24
Murdoch, Rupert, 5
Music, 1, 9-14, 16-17, 20, 28-31, 34-35, 39, 42-43, 75, 83, 85, 91-92, 94-96, 105, 119, 138, 151-152, 163, 182-183, 185-186, 220
Music Mixer, 11
Mute, 235-236

Mux, 149-150

**N**
NAB, 14
NBA, 146-147
NBA Championship, 146-147
NBC,
Olympics, 33, 130-131, 164
Saturday Night Live, 14
Nederlandsche Omroep Stichting, 192
NEP, 4
Net Weight, 235-236
NETL, 67
Neumann KM84, 37
Neve, Rupert, 51-52
New Jersey, 130-131
New York, 7-8, 130-131
NFL,
146, 158-159
Superbowl, 158-159
NHK, 37, 39, 42, 44
NHK Japan, 39, 44
No-Epoxy, 157
Nonmetallic, 218-219
Nonreversible, 158-159

**O**
Oaklink, 158-159
OB, 1, 4-6, 8-11, 14, 21, 39-40, 53-56, 85, 102, 104-105, 124, 132, 135, 145-146, 152, 156, 207
OB Van, 4, 8-11, 14, 21, 39-40, 54-56, 85, 102, 104, 124, 132, 135, 145-146, 152, 156, 207
OB-140, 51-52
Obtaining Permissions, iv
OBV, 45, 54, 161
Ocala I, 232-233
Ohlmeyer, Don, 1
Olympics, 16-17, 48, 177, 199, 205, x
Ceremonies, 48
Games, 177, 199

Omni, 168, 171, 174-175, 196

Omnidirectional Microphones, 7, 167-169, 174-176, 186, 196

Onand, 188-189, 201

On-axis, 24, 168, 175-176, 178, 188-189, 191-192

Opening Ceremonies of the Sydney Olympics, 138

Operation, 16, 19, 21, 49, 51-52, 88, 101, 113, 116, 119-121, 125, 135-136, 224, 234, 236

Operators, 9-10, 14, 23-24, 27-30, 49, 92, 100, 109, 114-115, 119-120, 124-125, 132, 136, 232-233

Optical, 59-60, 141-142, 145-146, 149-150, 157-159

Organizational, 10-11

Organize, 19-20, 75-76, 116, 132

Organizing, 2-3, 10-11, 19, 27, 95, 108, 182

Orlando, 230, 232-233, 236

ORTF, 192

Otari, 158-159

Otari Lightwinder, 158-159

Outboard, 29, 51, 59, 92

Outdoor, 5, 116-118, 200

Output, 33, 35, 42, 46, 49, 51, 57-62, 64-66, 73, 79, 81-83, 85, 89-93, 102, 114-115, 128-129, 136-139, 149-150, 153-154, 177-180, 182, 191, 196, 207-208, 222-223, 226, 234-236

Connectors Microphone, 235

Outputs, 2-3, 20-21, 46, 49, 51, 53, 56, 77, 79-81, 85-86, 90, 92, 95, 102-103, 126, 136-137, 141-146, 154-155, 158-159, 191, 199, 235-236

Outside, 1, 4, 7-8, 14, 16, 29, 99, 101, 114, 120-122, 125, 132, 145-148, 172, 184, 207, 224, 232

Overloading, 38, 65-66, 114-115, 191

Owens, Jim, 16-17

**P**

Packets, 130-132, 153-154, 156

Panorama, 57, 79, 198

PAP, 138

Parabolic, 205

Parametric, 65-68, 71, 89

Party-line, 120-122, 124-127, 135, 138

PAs, 220-221

Patch, 22, 80-81, 84, 136, 145-149

Patchbays, 21-22

Patching, 15, 21-22, 51, 80, 95-96, 147-148

Pattern Is Best, 174

PC, 95, 221, 235

PCC, 188

PCM, 128

Peak Program Meter, 64-65

PFL, 57, 85

Phantom, 57-59, 75, 153-154, 158-160, 181-182, 207, 214

Photo, 116, 124, x

Photo, 116, 124

Pickup, 14, 168-171, 173-174, 180-181, 188-189, 194, 198, 208, 228

Pilot, 235-236

Pin, 145-146, 181-182, 190, 207

PL, 101-102, 104, 106, 108-109, 114-116, 119, 123, 133, 160-161, 231-232

Planning, 15-16, 21, 80, 165-166

Plasma PPM, 65

Play, 20, 35-36, 99, 102, 111, 113, 117-118, 146-147, 172, 175, 185

Playback, 2, 7-8, 10-14, 19-21, 27, 39-42, 45-46, 51, 55, 58, 74-75, 85, 91-92, 94-96, 135, 183

Plexiglass, 203-204

PLII, 40-41, 44

PLL, 222

PLs—Communications, 108

Plug-in, 51-52, 213

Podium, 20, 170, 183, 203, 221

POF, 150

Point-to-point, 23-24, 109, 125-127, 133, 135, 145-146

Polar, 168-170, 173, 188-189, 191-192, 196

Polish, 157-158
  Polish–ST, 157
Port, 108-109, 128-130, 137-138
Post, 12, 26, 76, 116-117, 195
Post-fade, 75
Post-production, 2-3, 12
  Post-Production Sound, 2-3, 12
POV, 165-166, 185-186, 194
Power, 9-10, 16-17, 51-52, 57-61, 75, 104, 106,
  108, 120-121, 123-124, 126-128, 133, 136-
  138, 141, 143-144, 149-155, 157-161, 175-
  176, 179-182, 188-189, 207, 214, 216-217,
  227, 232, 234-235
Powered Antenna, 229
Powered Party Lines, 120-121
PPM, 42, 64-65
Precabled, 103, 145-146
  Precabled Venues, 103, 145-146
Prefade, 75, 85, 110
Preloaded Epoxy, 157-158
Pre-recorded, 42
Pressure Record Process, 188-189
Pro Logic II, 32, 35-36, 38, 40-41, 44, 46
Problems, 5-6, 14, 22, 28-29, 32, 34, 39-41, 46,
  56, 86, 90-91, 107, 110, 113-115, 120-121,
  136-137, 141, 145-146, 161, 170-171, 188-
  189, 199, 207, 224, 226, 234
  Problemsolver, 22
Processing Equipment, 9, 32, 91-92
Producer, 2, 10-11, 15-16, 19-21, 23-24, 27-28,
  37, 49, 75, 85, 91-92, 102, 105, 107-110,
  120-122, 128, 135-137, 182, 185, 220-221
Producers, 1, 16, 23-24, 28, 40-41, 100, 125,
  132
Production, 1-17, 19-20, 24, 27-30, 33, 39-42,
  45-46, 49, 51, 55, 74-75, 83-85, 91-92, 99-
  100, 102, 105-106, 108-111, 113-116, 119-
  121, 123, 125, 132, 163-165, 182-183, 185,
  195, 199, 201, 207, 220, 222, 231-232
  Plan, 20, 113
  Services, iv
Production-specific, 20, 28, 55

Professional Wireless Systems, 230, 236
Programmable, 21, 24, 50-51, 68, 71-72, 90, 94,
  108-109, 126-128, 130-131, 135-136, 235
Programming, 1-4, 10, 15, 22, 29-31, 42, 45, 95,
  112, 119, 132, 232-233
PRP, 188-189
Punch, 28, 31, 147-148, 155
Putnam, Bill, 51-52
Putting It Together, 207
PWS, 230
PZM, 188-189

**Q**
Quality, 5-11, 14-15, 20-21, 27-28, 30, 51-52,
  54, 104-105, 111-112, 131, 136, 138-139,
  141, 147, 153, 158-159, 164-166, 169, 179,
  190-192, 207, 225, 235, xi
Quest, 5

**R**
Racing, 4-6, 10-12, 20-21, 27, 114-115, 184,
  209
Rack of AT5000 Receivers, 223
Rack-mounted, 218-220
Radio Transmission, 216-217
Radio World Streicher, 209
Radio-frequency, 151-152, 224
  Interference, 151-152, 224
RCA, 4, 143-144
Rear, 29, 31-32, 34-39, 46, 169-172, 174-175,
  185, 228, 230
  Rear Left, 31-32, 185
  Rear Right, 31-32, 34, 185
Reassignment, 73
Receiver, 110, 126, 138-139, 149-150, 158, 211-
  212, 218-227, 229-232, 235-236
Receivers, 31, 138-140, 216-221, 223-224, 226,
  231-232, 235
  In FM, 218-219
Recorded Sound, 51-52
Recording, 2, 8-9, 11, 14, 16-17, 20, 22, 27, 29-
  30, 35, 45-46, 51-53, 64, 74-75, 77-78, 94,

132, 153-154, 163, 174, 182, 188-189, 191, 195, 197-198, 209

Recreating, 183

Reduction, 68, 71, 92, 189-190, 230-231

References, 60, 63-64, 87, 209, ix

Refining, 5-6

Reflected, 149-150, 169, 174, 178, 188-189, 207, 218-219, 224-225, 230

Reflections, 207, 218-219, 227

Reflective, 168-169, 174-175, 187, 189-191, 207, 218-219

Reidel, 158-159, 231-232

Relays, 2, 126-127, 130-131

Remote, 1-2, 4-10, 14, 16-17, 51-52, 59, 65-66, 94-95, 101-102, 104, 114-115, 119, 122, 125, 129-131, 136-137, 142, 145-146, 148, 153-154, 158-160, 190, 220, 224

Replay, 20, 24, 27, 45-46

Reproduction, 5, 11, 27-29, 40-41, 71-72, 163, 165-166, 170, 177-178, 182, 185, 199

Research, 149-150, 236

Research Fox, 236

Reverberant Field, 207

RF, 9-11, 14, 45, 139-140, 142, 211-212, 218-219, 221, 224, 226-227, 230-236

    Audio, 14

    PL, 231-232

RG-58, 230

RG-59, 230-231

RG-8, 230-231

Ryan, Dennis, 7, 16, 31, 95-96, 164

Riedel, 106, 130-131

    Riedel Communications, 130-131

Rigging, 9-10, 175

Right, 7, 25-26, 29-36, 40-42, 44, 46, 64, 79-80, 82, 84-86, 89, 93, 95-96, 114, 135-137, 169-170, 184-185, 191, 194-195, 197-199, 223, 230

    Right Total, 32, 44

    Right Total, 32, 44

    Rights, 4, 10, 113, iv

RJ-12, 129

RJ-45, 131, 235

RJ-45 Ethernet, 131

RMS, 61-62, 64, 96-97

Robertson, Chad, 15

ROH Systems, 120-121

Room Acoustics, 207

    Direct Sound, 207

Romero, Manolo, 16-17

Rotate, 158

Rotating, 73

Routing, 2-3, 5-6, 21-22, 33, 40-41, 49-52, 54, 57, 59-60, 73-74, 79-80, 82, 86, 88-90, 130-131, 153-155

Roving, 99, 231-232

RS-422, 154-155

RS-485, 129

RTS ADAM, 129, 231-232

    RTS Intercom Systems, 124

    RTS PS, 124

    RTS Systems, 120-121, 123-125, 132, 135, 137

RVON-8, 131-132

RVON-C, 131

Rycote, 112, 200

**S**

Safe Microphone Placement, 175-176

Safety, 9-10, 24, 100, 110, 138-139, 145, 158-159, 172, 175-176, 186

Saltans, Peteris, 22

Samplers, 14

Saturday Night Live, 13-14

SBE, 232-233

SC, 157-158

Scalise, Ron, 15-16, 45, 145, 164

Scenes, 95

Scotch Super, 234-235

Screens, 21, 49-50, 56, 63-65, 73, 87, 92, 172, 199

SCSI, 158-159

Sculpting, 28-30, 40-41

SD, 91

SDI, 141, 153-154

SDTV, 158-159

Senior Audio, 2-3, 9-11, 15-16, 19-20, 22, 27-28, 99-101, vii, xi-xii

Sennheiser, 37, 196, 201, 209, 222, 235-236, ix
EM550G2 Wireless Twin Receiver, 235-236

Seoul Olympics, 126-127

Separation, 42, 186-187, 191-192, 227

Setting, 15, 69, 76, 85, 91-92, 103, 113, 216-217, 224

Set-up, 2-3, 7-11, 16, 19-21, 24-25, 27-29, 33, 35-36, 40-41, 56, 59, 63, 77-78, 99-108, 132-133, 157, 224, xi-xii

Shaping, 15-16, 65-66

Shelving EQs, 68, 71-72

Short Shotgun, 26, 192

Short Stereo Shotgun, 26

Shotgun Microphones, 14, 35-36, 163-164, 172-173, 175-176, 181, 186-187, 190-192, 195, 207
Shotguns, 111, 171, 186, 190-191, 194

Shure, 22-23, 209, 222, 236

Side-line Reporters, 110

Seiderman, Bob, 5-6, 16-17, 53

Signal,
Delay, 156
Path, 5-6, 38, 54, 56-57, 64-66, 71, 74, 79, 81, 143-144, 207, 227
Signal Path A, 56, 79

Signals, 3, 10-11, 16-17, 21, 31-32, 40-41, 49, 51-52, 55, 58-59, 61, 63-64, 73-75, 84-85, 89, 91, 95, 104-105, 114-116, 125-127, 129, 135-136, 138-139, 141-146, 149-151, 153-156, 158-161, 189-190, 195, 197-198, 207, 216-219, 222-225

Single-ear, 135-136

Single-mode, 150-151

Single-tube, 196

Sixteen-channel, 108

Size, 7-8, 19, 21, 54-56, 85, 90, 119, 126, 129, 131-132, 141, 156, 164, 169, 178, 184-186, 188-189, 192-193, 203, 205, 208, 218-219

Small, 5-6, 9-10, 14-15, 23, 51-52, 54, 56-57, 59, 63, 75-77, 85, 90, 96-97, 102-103, 105, 114, 120-121, 141, 151, 156, 168, 175, 177, 180-181, 184, 189-190, 207, 213, 220-221
Small-diaphragm, 174-176, 186
Small-diaphragm Microphones, 174-176, 186

Smaller TV, 125

SNR, 224

Society of Broadcast Engineers, 232-233

Solder, 114-115

Solid State Logic, 51-52, 55, 57, 71

Solo, 57, 71-72, 85, 89
Solo/PFL, 57

Sony, 201

Sound,
Designer, 2-3, 15-16, 24, 27-28, 31, 33, 91-92, 182, 198-199, 208
Elements, 15-16, 20-21, 24, 27-28, 30, 34, 40-42, 83, 163, 182
Mix, 5-6, 10-11, 20, 23-24, 28-29, 31-32, 34, 39-42, 47, 84, 91-92, 105, 125, 184, 186, 198-199, 208, 220-221, xi
Recordist, 14, 27, 195

Soundcheck, 236

Soundcraft, 53

Soundfield, 198

Sounds, 14, 28-31, 34-36, 40, 46, 55, 71-72, 78, 80, 163, 168, 170, 174-175, 180, 183, 186-187, 190, 199, 208, 224, xi-xii

Soundtrax, 56

Spatial, 5, 7, 28-30, 40-41, 80, 165-166, 184-186, 191-192, 194, 198, 208
Orientation, 29, 80, 165-166, 198, 208

Speaker, 2-3, 5-6, 8, 28-32, 34, 38-41, 46, 56, 80, 83-85, 95, 105, 110, 113-114, 120-121, 135-137, 185, 221, xi-xii
Monitoring, 5-6, 85, 95

Specialty Microphones, 111, 186, 200-201

Spectrum Allocation, 232-233

SPLs, 60, 63, 191, 175-176, 182

Sports, 1-5, 7-10, 13, 16-17, 20, 30-31, 33-35,

40-42, 45-46, 54-55, 74, 83, 93-95, 99, 105, 110-116, 119, 132, 136-137, 141-142, 163, 175-176, 183-186, 188-190, 200-201, 207, 211, 214, 231-232
Spotter PL, 102
Square, 61, 207
SRS, 40-41
SSA-424, 135
SSL, 51-52, 54-55, 57, 68-69, 77
   SSL200, 64
ST, 150, 157-158
Stage Manager PL, 102
Staples Center, 147-148
Statistician PL, 102
Stereo Microphones, 25, 29-31, 35, 189-190, 192, 195-196
Stereo Recording, 191, 195, 197-198, 209
Stereo XY, 26, 35, 196
Stereo-effects, 25
Stereophonic TV, 209, ix
Stickiness, 96-97
STL, 234
Stoffo, James, 230, 232-233, 236
Store Peak, 65
Stories, 5, 12-13, 39, 175-176
Studio Control Room Productions, 12
Subgrouping, 77
Sub-Mixer, 10-11, 22, 118
Summer Games, 164, 175-176
Summer Olympics, 38-39, 46
Super Bowl, 166, 230
Super Shotgun, 26
Super-directional DSP, 183
Superstation Channel, 4
Support, 7-8, 20, 31, 100, 105, 108, 123, 132, 141-142
Surface-mount Microphone, 187-188
Surround, 1-2, 5-8, 25-26, 29-44, 46, 49, 51-52, 55, 80-81, 83-85, 90-94, 112, 164-166, 177-178, 183, 185, 191, 197-199, 208-209
   Formats, 31, 43-44
   Sound, 1-2, 5-8, 31, 33, 38-40, 42, 51-52,

80-81, 90, 94, 164-166, 185, 191, 197-198, 209
Sound Microphone Techniques, 198
Stereo, 30-31, 35-37, 39-41, 43-44, 49, 84, 93, 183, 191, 199, 208
Surround-sound, 5-6, 20, 29, 31-34, 39-42, 46, 62, 81, 85, 184-185, 198-199
Switches, 71-72, 79, 91, 96-97, 126, 156, 216, 225, 227
Sydney, 138, 158-159, 164, 220-221
   Olympics, 138, 158-159, 220-221
System Connectivity, 145
System Design, 132
Systems, 9, 20, 23-24, 27, 31-32, 40-41, 47, 51-52, 103-106, 109, 113-115, 119-121, 123-132, 135-141, 145-146, 150-155, 158-161, 207, 211, 218-222, 224, 227, 230-233, 235-236
Systems Review, 235

**T**
T1, 119, 130-131
Talkback, 89, 102, 105, 133, 136-137, 220-221, 232
Talladega Motor Speedway, 184
Tape, 2-3, 14, 20, 53-54, 75, 100, 109, 116-117, 124, 144, 234-235
TC, 42-44, 94
TC Electronics, 42-44, 94
   TC6000, 42
TDM, 127-129
Technicians, 1, 5, 8-10, 15-17, 43, 60, 100, 125, 132, 148
Techniques, 1, 24, 47, 174-175, 184, 191-192, 194-198, 209
Technology, 1-3, 5-6, 8, 15-17, 24, 31, 40-41, 90, 116, 120-121, 124, 126-128, 130-131, 138-139, 145-146, 149-150, 153-154, 165, 211
Telco, 147, 149-150
Telecast Adder System, 158-159
Telecaster, 30

Telephone, 106, 108-109, 116, 118-121, 124-127, 130-131, 147-148
    Telephone Hybrids, 124-125
Television Communications Jargon, 102
Telex, 106, 108, 130-132, ix
    Intelligent Trunking, 132
Tennis, 38-39, 175, 185, 195
The Ed Sullivan Show, 1
The Holophone, 199
The Modern Announce Booth, 132-133
Third-order, 224
Three-camera, 9
Three-position, 235
Threshold, 63, 72-73, 92, 132
Time Division Multiplexing, 127
TLM, 37
Tone Lock, 235
Torino, 130-131
Total Harmonic Distortion, 235
Touch, 50, 87, 96-97, 116-117, 139-140, 158-159, 207
    Touch-sensitive, 50, 87
Touring, 53, 125
Translating, 20
Transmission, 5-6, 27-29, 31, 38, 40-41, 49, 83-84, 126-127, 133, 150-151, 154-155, 157, 199, 216-217, 234
Transmitter, 110, 138-140, 149-150, 158, 211-223, 227, 230-232, 234-236
TRIMS, 138
Troubleshooting, 2-3, 16, 21-22, 103-105, 109, 113-116, 145-146
    Communications Cables, 114-115
    Microphone Cables, 114
TRS, 235
Trunking, 130, 132
TS, 235
Tuned, 138-140, 198-199, 227, 231-232
Turner, Ted, 4
    Turner Television, 51-52
TV, 1, 4-5, 9-10, 120-122, 125, 130-131, 136-137, 209, 222, 232-234, vii, ix

Sound Engineering, 1
    Channel, 222
TW, 135
Two-wire, 124-125, 128, 130-131, 133, 135-137, 161

**U**
UHF, 227, 229-235
    High, 234
    Low, 234
    Mid, 234
    Wireless, 224
Unbalanced, 135-136, 143-145, 208, 235
Up-mixing, 42-43
    Up-Mixing Surround, 42
UPS, 161
USA Today, 1
US Patent, 188-189
Useful Tricks, 195
User Stations, 108, 119-121, 124-126, 128-131, 133, 135
Utilities, 16

**V**
Vacuum, 2, 51-52, 96-97
VAD, 132
Vadis, 153-154, 158-159
Variation, 183, 194, 197-198, 216-218
Variety of Formats, 30
VCA, 76-79
    Grouping, 76-77
Veledrome, 203
Venue PA, 102
VHF, 229-231, 234
Video-based, 158-159
Viewers, 2, 23, 28, 31-32, 184
Voice, 7-8, 28-31, 34-35, 42, 83, 95-96, 99, 105, 110, 126-127, 130-132, 136, 139-140, 181, 195, 203
VOIP, 119, 130-131
Voltage RMS, 62
VU, 61-65, 69

**W**

Ward Beck Mixing Desk, 5-6, 54
Water, 117, 203, 207, 235
Waterproofing, 117, 234-235
    RF, 234-235
Wells, Mike, 1
Wet, 113, 143-144, 200, 203
WGN Chicago, 4
Whirlwind, 154-155
Wickersham, Ron, 188-189
Wimbledon, 38-39
Wind, 38, 69, 116, 167, 191, 199, 207
Wind Noise, 38, 116, 167, 207
Wind Zeppelins, 199
Windows, 87, 221
Windscreens, 199
Winter Olympics, 38-39
Wire, 9, 21, 44, 60, 80, 101, 104-105, 113-118,
    124-126, 129, 133, 141-152, 161, 178, 190,
    207-208
Wireless,
    Communications, 220-221, 231-232
    IFB, 139-140, 231-232
    Intercom Systems, 138-140, 231-232
    Microphones, 1, 14, 20, 91, 138-140, 183,
    211-212, 214-215, 218-222, 224, 231-232,
    234, 236

    PL, 231-232
    Systems, 138-140, 211, 218-219, 222, 224,
    227, 230-232, 236, vii
Workplace, 7-8
World Cup, 158-159, 205
    Football, 10
    Skiing, 158-159
World Series, 146-147
Wrap, 51, 117, 142, 234-235

**X**

X Games, 130-131, 145
XLR, 102, 143-146, 207
    XLR-3, 143-144
    XLRM-type Instrument, 235
    XLR-type, 181-182
XY, 26, 35, 191-193, 196

**Y**

Yagi, 229-230
Yamaha, 53, 72, 82, 158-159, ix

**Z**

Zeppelin-style, 200